Mathematical Logic

OXFORD TEXTS IN LOGIC

Mathematical Logic

IAN CHISWELL and WILFRID HODGES

OXFORD

UNIVERSITY PRESS

OXFORD

UNIVERSITY PRESS

Great Clarendon Street, Oxford OX2 6DP
United Kingdom

Oxford University Press is a department of the University of Oxford.
It furthers the University's objective of excellence in research, scholarship,
and education by publishing worldwide. Oxford is a registered trade mark of
Oxford University Press in the UK and in certain other countries

© Ian Chiswell and Wilfrid Hodges, 2007

The moral rights of the author have been asserted

First published 2007
Reprinted 2013

Published in the United States of America by Oxford University Press
198 Madison Avenue, New York, NY 10016, United States of America

British Library Cataloguing in Publication Data
Data available

Library of Congress Cataloging in Publication Data
Data available

ISBN 978-0-19-921562-1

Preface

This course in Mathematical Logic reflects a third-year undergraduate module that has been taught for a couple of decades at Queen Mary, University of London. Both the authors have taught it (though never together). Many years ago the first author put together a set of lecture notes broadly related to Dirk van Dalen's excellent text *Logic and Structure* (Springer-Verlag, 1980). The present text is based on those notes as a template, but everything has been rewritten with some changes of perspective. Nearly all of the text, and a fair number of the exercises, have been tested in the classroom by one or other of the authors.

The book covers a standard syllabus in propositional and predicate logic. A teacher could use it to follow a geodesic path from truth tables to the Completeness Theorem. Teachers who are willing to follow our choice of examples from diophantine arithmetic (and are prepared to take on trust Matiyasevich's analysis of diophantine relations) should find, as we did, that Gödel's Incompleteness Theorem and the undecidability of predicate logic fall out with almost no extra work. Sometimes the course at Queen Mary has finished with some applications of the Compactness Theorem, and we have included this material too.

We aimed to meet the following conditions, probably not quite compatible:

- The mathematics should be clean, direct and correct.

- As each notion is introduced, the students should be given something relevant that they can do with it, preferably at least a calculation. (For example, parsing trees, besides supporting an account of denotational semantics, seem to help students to make computations both in syntax and in semantics.)

- Appropriate links should be made to other areas in which mathematical logic is becoming important, for example, computer science, linguistics and cognitive science (though we have not explored links to philosophical logic).

- We try to take into account the needs of students and teachers who prefer a formal treatment, as well as those who prefer an intuitive one.

We use the Hintikka model construction rather than the more traditional Henkin-Rasiowa-Sikorski one. We do this because it is more hands-on: it allows us to set up the construction by deciding what needs to be done and then doing it, rather than checking that a piece of magic does the work for us.

We do not assume that our students have studied any logic before (though in practice most will at least have seen a truth table). Until the more specialist matter near the end of the book, the set theory is very light, and we aim to explain any symbolism that might cause puzzlement. There are several proofs by induction and definitions by recursion; we aim to set these out in a format that students can copy even if they are not confident with the underlying ideas.

Other lecturers have taught the Queen Mary module. Two who have certainly influenced us (though they were not directly involved in the writing of this book) were Stephen Donkin and Thomas Müller—our thanks to them. We also thank Lev Beklemishev, Ina Ehrenfeucht, Jaakko Hintikka, Yuri Matiyasevich and Zbigniew Ras for their kind help and permissions with the photographs of Anatoliĭ Mal'tsev, Alfred Tarski, Hintikka, Matiyasevich and Helena Rasiowa respectively. Every reasonable effort has been made to acknowledge copyright where appropriate. If notified, the publisher will be pleased to rectify any errors or omissions at the earliest opportunity.

We have set up a web page at
www.maths.qmul.ac.uk/~wilfrid/mathlogic.html
for errata and addenda to this text.

Ian Chiswell
Wilfrid Hodges
School of Mathematical Sciences
Queen Mary, University of London
August 2006

Contents

1 Prelude

1.1 What is mathematics?

Euclid Egypt, c. 325–265 BC.
For Euclid, mathematics consists of proofs and
constructions.

Al-Khwārizmī Baghdad, c. 780–850.
For Al-Khwārizmī, mathematics consists of
calculations.

G. W. Leibniz Germany, 1646–1716.
According to Leibniz, we can calculate whether a
proof is correct. This will need a suitable language
(a *universal characteristic*) for writing proofs.

Gottlob Frege Germany, 1848–1925.
Frege invented a universal characteristic. He called
it *Concept-script* (Begriffsschrift).

Gerhard Gentzen Germany, 1909–1945.
Gentzen's system of *natural deduction* allows us to
write proofs in a way that is mathematically
natural.

1.2 Pronunciation guide

To get control of a branch of mathematics, you need to be able to speak it. Here are some symbols that you will probably need to pronounce, with some suggested pronunciations:

$\bot$	'absurdity'
$\vdash$	'turnstile'
$\models$	'models'
$\forall$	'for all'
$\exists$	'there is'
t_A	'the interpretation of t in A'
$\models_A \phi$	'A is a model of ϕ'
$\approx$	'has the same cardinality as'
$\prec$	'has smaller cardinality than'

The expression '$x \mapsto y$' is read as 'x maps to y', and is used for describing functions. For example, '$x \mapsto x^2$' describes the function 'square', and '$n \mapsto n+2$' describes the function 'plus two'. This notation is always a shorthand; the surrounding context must make clear where the x or n comes from.

The notation '$A \Rightarrow B$' is shorthand for 'If A then B', or 'A implies B', or sometimes 'the implication from A to B', as best suits the context. Do not confuse it with the notation '$\rightarrow$'. From Chapter 3 onwards, the symbol '$\rightarrow$' is not shorthand; it is an expression of our formal languages. The safest way of reading it is probably just 'arrow' (though in Chapters 2 and 3 we will discuss its translation into English).

The notation '$\mathbb{N}$' can be read as 'the set of natural numbers' or as 'the natural number structure', whichever makes better sense in context. (See Example 5.5.1 for the natural number structure. Note that our natural numbers are $0, 1, 2, \ldots$, starting at 0 rather than 1.)

The following rough pronunciations of personal names may help, though they are no substitute for guidance from a native speaker:

Frege: *FRAY-ga*

Hintikka: *HIN-ticka*

Leibniz: *LIBE-nits*

Łoś: *WASH*

Łukasiewicz: *woo-ka-SHAY-vitch*

Matiyasevich: *ma-ti-ya-SAY-vitch*

Giuseppe Peano: *ju-SEP-pe pay-AH-no*

Peirce: *PURSE*

Helena Rasiowa: *he-LAY-na ra-SHOW-va*

Scholz: *SHOLTS*

Dana Scott: *DAY-na SCOTT*

Sikorski: *shi-COR-ski*

Van Dalen: *fan DAH-len*

Zermelo: *tser-MAY-low*

2 Informal natural deduction

In this course we shall study some ways of proving statements. Of course not every statement can be proved; so we need to analyse the statements before we prove them. Within propositional logic we analyse complex statements down into shorter statements. Later chapters will analyse statements into smaller expressions too, but the smaller expressions need not be statements.

What is a statement? Here is a test. A string S of one or more words or symbols is a statement if it makes sense to put S in place of the '...' in the question

Is it true that ...?

For example, it makes sense to ask any of the questions

Is it true that π is rational?
Is it true that differentiable functions are continuous?
Is it true that $f(x) > g(y)$?

So all of the following are statements:

π is rational.
Differentiable functions are continuous.
$f(x) > g(y)$.

For this test it does not matter that the answers to the three questions are different:

No.
Yes.
It depends on what f, g, x and y are.

On the other hand, none of the following questions make sense:

Is it true that π?
Is it true that Pythagoras' Theorem?
Is it true that $3 + \cos\theta$?

So none of the expressions 'π', 'Pythagoras' Theorem' and '$3 + \cos\theta$' is a statement.

The above test assumes that we know what counts as a 'symbol'. In practice, we do know and a precise definition is hardly called for. But we will take for granted (1) that a symbol can be written on a page—given enough paper, ink, time and patience; (2) that we know what counts as a finite string of symbols; (3) that any set of symbols that we use can be listed, say as $s_0, s_1, s_2, \ldots$, indexed by natural numbers. In some more advanced applications of logic it is necessary to call on a more abstract notion of symbol; we will discuss this briefly in Section 7.9.

2.1 Proofs and sequents

Definition 2.1.1 A mathematical proof is a proof of a statement; this statement is called the *conclusion* of the proof. The proof may use some assumptions that it takes for granted. These are called its *assumptions*. A proof is said to be a proof *of* its conclusion *from* its assumptions.

For example, here is a proof from a textbook of pure mathematics:

Proposition   Let $z = r(\cos\theta + i\sin\theta)$, and let n be a positive integer. Then

$$z^n = r^n(\cos n\theta + i\sin n\theta).$$

Proof   Applying Theorem 6.1 with $z_1 = z_2 = z$ gives

$$z^2 = zz = rr(\cos(\theta + \theta) + i\sin(\theta + \theta)) = r^2(\cos 2\theta + i\sin 2\theta).$$

Repeating, we get

$$z^n = r \cdots r(\cos(\theta + \cdots + \theta) + i\sin(\theta + \cdots + \theta)) = r^n(\cos n\theta + i\sin n\theta). \quad \square$$

The proof is a proof of the equation

$$(2.1) \qquad\qquad z^n = r^n(\cos n\theta + i\sin n\theta)$$

so this equation (2.1) is the conclusion of the proof. (Note that the conclusion need not come at the end!) There are several assumptions:

- One assumption is stated at the beginning of the proposition, namely

$$z = r(\cos\theta + i\sin\theta), \text{and } n \text{ is a positive integer.}$$

  (The word 'Let' at the beginning of the proposition is a sign that what follows is an assumption.)

- Another assumption is an earlier theorem, mentioned by name:
     Theorem 6.1.

  (Note that this assumption is referred to but not written out as a statement.)

- Finally, there are a number of unstated assumptions about how to do arithmetic. For example, the proof assumes that if $a = b$ and $b = c$ then $a = c$. These assumptions are unstated because they can be taken for granted between reader and writer.

When we use the tools of logic to analyse a proof, we usually need to write down statements that express the conclusion and all the assumptions, including unstated assumptions.

A proof P of a conclusion ψ need not show that ψ is true. All it shows is that ψ is true *if* the assumptions of P are true. If we want to use P to show that ψ is true, we need to account for these assumptions. There are several ways of doing this. One is to show that an assumption says something that we can agree is true without needing argument. For example, we need no argument to see that $0 = 0$.

A second way of dealing with an assumption is to find another proof Q that shows the assumption must be true. In this case the assumption is called a *lemma* for the proof P. The assumption no longer counts as an assumption of the longer proof consisting of P together with Q.

Section 2.4 will introduce us to a third and very important way of dealing with assumptions, namely to *discharge* them; a discharged assumption is no longer needed for the conclusion. We will see that—just as with adding a proof of a lemma—discharging an assumption of a proof will always involve putting the proof inside a larger proof. So mathematical proofs with assumptions are really pieces that are available to be fitted into larger proofs, like bricks in a construction kit.

Sequents

Definition 2.1.2 A *sequent* is an expression

$$(\Gamma \vdash \psi) \quad (\text{or } \Gamma \vdash \psi \text{ when there is no ambiguity})$$

where ψ is a statement (the *conclusion* of the sequent) and Γ is a set of statements (the *assumptions* of the sequent). We read the sequent as 'Γ entails ψ'. The sequent $(\Gamma \vdash \psi)$ means

(2.2) There is a proof whose conclusion is ψ and whose undischarged assumptions are all in the set Γ.

When (2.2) is true, we say that the sequent is *correct*. The set Γ can be empty, in which case we write $(\vdash \psi)$ (read 'turnstile ψ'); this sequent is correct if and only if there is a proof of ψ with no undischarged assumptions.

We can write down properties that sequents ought to have. For example:

SEQUENT RULE (Axiom Rule) If $\psi \in \Gamma$ then the sequent $(\Gamma \vdash \psi)$ is correct.

SEQUENT RULE (Transitive Rule) If $(\Delta \vdash \psi)$ is correct and for every δ in Δ, $(\Gamma \vdash \delta)$ is correct, then $(\Gamma \vdash \psi)$ is correct.

Sequent rules like these will be at the heart of this course. But side by side with them, we will introduce other rules called *natural deduction rules*. The main difference will be that sequent rules are about provability in general, whereas natural deduction rules tell us how we can build proofs of a particular kind (called *derivations*) for the relevant sequents. These derivations, together with the rules for using them, form the *natural deduction calculus*. In later chapters we will redefine sequents so that they refer only to provability by natural deduction derivations within the natural deduction calculus. This will have the result that the sequent rules will become provable consequences of the natural deduction rules. (See Appendix A for a list of all our natural deduction rules.)

Derivations are always written so that their conclusion is their bottom line. A derivation with conclusion ϕ is said to be a *derivation of* ϕ.

We can give one natural deduction rule straight away. It tells us how to write down derivations to justify the Axiom Rule for sequents.

NATURAL DEDUCTION RULE (Axiom Rule) Let ϕ be a statement. Then

$$\phi$$

is a derivation. Its conclusion is ϕ, and it has one undischarged assumption, namely ϕ.

Both sequent rules and natural deduction rules were introduced in 1934 by Gerhard Gentzen as proof calculi. (A *proof calculus* is a system of mathematical rules for proving theorems. See Section 3.9 for some general remarks about proof calculi.) Gentzen's sequents were more complicated than ours—he allowed sets of statements on the right-hand side as well as the left. His *sequent calculus* lies behind some other well-known proof calculi such as tableaux (truth trees), but we will not study it in this course.

Exercises

2.1.1. What are the conclusion and the assumptions of the following argument?

Theorem Let r be a positive real number. Then r has a square root.

Proof Write $f(x) = x^2 - r$ for any real x. Then f is a continuous function on $\mathbb{R}$. If $x = 0$ then $f(x) = 0 - r < 0$ since r is positive. If x is very large then $f(x) = x^2 - r > 0$. So by the Intermediate Value Theorem there must be x such that $f(x) = 0$. For this value x,

$$r = r + 0 = r + f(x) = r + (x^2 - r) = x^2.$$

2.1.2. A first-year calculus textbook contains the following paragraph:

Given that

$$1 - \frac{x^2}{4} \leqslant u(x) \leqslant 1 + \frac{x^2}{2} \quad \text{for all } x \neq 0,$$

we calculate $\lim_{x \to 0} u(x)$. Since

$$\lim_{x \to 0} (1 - (x^2/4)) = 1 \quad \text{and} \quad \lim_{x \to 0} (1 + (x^2/2)) = 1,$$

the Sandwich Theorem implies that $\lim_{x \to 0} u(x) = 1$.

What is the conclusion of this argument, and what are the assumptions? (You can answer this question without knowing what all the expressions in it mean.)

2.1.3. From your understanding of mathematical arguments, which (if any) of the following possible sequent rules seem to be true? Give reasons.
Possible sequent rule A: If the sequent $(\Gamma \vdash \psi)$ is a correct sequent, and every statement in Γ is also in Δ, then the sequent $(\Delta \vdash \psi)$ is also correct.

Possible sequent rule B: If the sequent $(\{\phi\} \vdash \psi)$ is correct, then so is the sequent $(\{\psi\} \vdash \phi)$.

Possible sequent rule C: If the sequents $(\Gamma \vdash \psi)$ and $(\Delta \vdash \psi)$ are both correct, then so is the sequent $((\Gamma \cap \Delta) \vdash \psi)$.

2.2 Arguments introducing 'and'

Tell me about the word 'and' and its behaviour. Your knowledge of logic won't do you any good if you don't know about this word. (*Abu Saʿīd As-Sīrāfī*, AD *941*)

We shall study how the word 'and' appears in arguments. We are mainly interested in this word where it appears between two statements, as, for example, in

(2.3) v is a vector and α is a scalar.

We shall write this sentence as

(2.4)   (v is a vector $\wedge$ α is a scalar)

We shall write $\wedge$ for 'and' (between statements). The parentheses are an essential part of the notation; later we will adopt some rules for leaving them out, but only in contexts where they can be reconstructed.

There are a number of things that we can say with the help of this notation, even when the word 'and' does not appear between statements in the original English. Here are some typical examples:

(2.5)   The function f is surjective and differentiable
          (The function f is surjective $\wedge$ the function f is differentiable)

(2.6) $2 < \sqrt{5} < 3$
                          $(2 < \sqrt{5} \wedge \sqrt{5} < 3)$

The next example is a little subtler. If A and B are sets, then a necessary and sufficient condition for A and B to be the same set is that every member of A is a member of B (in symbols $A \subseteq B$) and every member of B is a member of A (in symbols $B \subseteq A$). So we can use our new symbol:

(2.7) $A = B$
                          $(A \subseteq B \wedge B \subseteq A)$

This paraphrase is often useful in proofs about sets.

Now consider how we prove a statement made by joining together two other statements with 'and'.

Example 2.2.1 We prove that $2 < \sqrt{5} < 3$. (Compare with (2.6).)

(1) We prove that $2 < \sqrt{5}$ as follows. We know that $4 < 5$. Taking positive square roots of these positive numbers, $2 = \sqrt{4} < \sqrt{5}$.

(2) We prove that $\sqrt{5} < 3$ as follows. We know that $5 < 9$. Taking positive square roots of these positive numbers, $\sqrt{5} < \sqrt{9} = 3$.

The moral of this example is that if we put together a proof of ϕ and a proof of ψ, the result is a proof of $(\phi \wedge \psi)$. The assumptions of this proof of $(\phi \wedge \psi)$ consist of the assumptions of the proof of ϕ together with the assumptions of the proof of ψ. We can write this fact down as a sequent rule:

SEQUENT RULE ($\wedge$I)   If $(\Gamma \vdash \phi)$ and $(\Delta \vdash \psi)$ are correct sequents then $(\Gamma \cup \Delta \vdash (\phi \wedge \psi))$ is a correct sequent.

The name ($\wedge$I) expresses that this is a rule about $\wedge$, and the symbol $\wedge$ is introduced (hence 'I') in the last sequent of the rule. We refer to this rule as $\wedge$-introduction.

We also adopt a schematic notation for combining the proofs of ϕ and ψ:

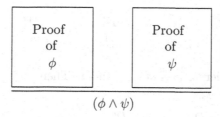

$$(\phi \wedge \psi)$$

This diagram represents a proof of $(\phi \wedge \psi)$, which appears at the bottom. This bottom expression is called the *conclusion* of the proof.

The box notation is a little heavy, so we adopt a lighter version. We write

(2.8)
$$\begin{array}{c} D \\ \phi \end{array}$$

to stand for a proof D whose conclusion is ϕ. Using this notation, we recast the picture above as a rule for forming proofs. This new rule will be our second natural deduction rule. We give it the same label ($\wedge$I) as the corresponding sequent rule above.

NATURAL DEDUCTION RULE ($\wedge$I)   If

$$\begin{array}{c} D \\ \phi \end{array} \quad \text{and} \quad \begin{array}{c} D' \\ \psi \end{array}$$

are derivations of ϕ and ψ respectively, then

$$\frac{\begin{array}{cc} \begin{array}{c} D \\ \phi \end{array} & \begin{array}{c} D' \\ \psi \end{array} \end{array}}{(\phi \wedge \psi)} \ (\wedge\text{I})$$

is a derivation of $(\phi \wedge \psi)$. Its undischarged assumptions are those of D together with those of D'.

Example 2.2.2 Suppose

$$\begin{array}{c} D \\ \phi \end{array}$$

is a derivation of ϕ. Then

$$\frac{\begin{array}{cc} \begin{array}{c} D \\ \phi \end{array} & \begin{array}{c} D \\ \phi \end{array} \end{array}}{(\phi \wedge \phi)} \ (\wedge\text{I})$$

is a derivation of $(\phi \wedge \phi)$. Its undischarged assumptions are those of D.

Example 2.2.3 Suppose

$$\begin{array}{ccc} D & D' & D'' \\ \phi & \psi & \chi \end{array} \text{ and }$$

are respectively derivations of ϕ, ψ and χ. Then

$$\cfrac{\cfrac{\begin{array}{cc} D & D' \\ \phi & \psi \end{array}}{(\phi \wedge \psi)}\ (\wedge\text{I}) \qquad \begin{array}{c} D'' \\ \chi \end{array}}{((\phi \wedge \psi) \wedge \chi)}\ (\wedge\text{I})$$

is a derivation of $((\phi \wedge \psi) \wedge \chi)$, got by applying $\wedge$-introduction twice; the second time we apply it with D'' as the second derivation. The undischarged assumptions of this derivation are those of D, those of D' and those of D''.

Remark 2.2.4 The following points will apply (with obvious adjustments) to all future derivations too.

- The conclusion of a derivation is the statement written in the bottom line.
- If the conclusion of an application of $(\wedge\text{I})$ is $(\phi \wedge \psi)$, then the derivation of ϕ must go on the left and the derivation of ψ on the right.
- In Example 2.2.2 we used the same derivation of ϕ twice. So the derivation must be written twice.
- As we go upwards in a derivation, it may branch. The derivation in Example 2.2.3 has at least three branches (maybe more, depending on what branches the derivations D, D' and D'' have). The branches stay separate as we go up them; they never join up again. A derivation never branches downwards.
- The name of the rule used in the last step of a derivation is written at the right-hand side of the horizontal line above the conclusion of the derivation. In our formal definition of derivations (Definition 3.4.1) these rule labels will be essential parts of a derivation.

Now by the Axiom Rule for natural deduction (Section 2.1), ϕ by itself is a derivation of ϕ with undischarged assumption ϕ. So in Example 2.2.3 the derivation D could be this derivation, and then there is no need to write 'D'. Similarly we can leave out 'D''' and 'D'''', regarding ψ and χ as derivations with themselves as conclusions. The result is the derivation

$$
(2.9) \qquad \dfrac{\dfrac{\phi \qquad \psi}{(\phi \wedge \psi)} \ (\wedge \mathrm{I}) \qquad \chi}{((\phi \wedge \psi) \wedge \chi)} \ (\wedge \mathrm{I})
$$

Now the undischarged assumptions of this derivation are those of D, D' and D'' together; so they are ϕ, ψ and χ. Thus the derivation (2.9) shows that there is a proof of $((\phi \wedge \psi) \wedge \chi)$ with undischarged assumptions ϕ, ψ and χ. In other words, it shows that the following sequent is correct:

$$
(2.10) \qquad \{\phi, \psi, \chi\} \vdash ((\phi \wedge \psi) \wedge \chi).
$$

Likewise if we cut out the symbol 'D' from Example 2.2.2, what remains is a derivation of $(\phi \wedge \phi)$ from ϕ and ϕ, establishing the correctness of

$$
(2.11) \qquad \{\phi\} \vdash (\phi \wedge \phi).
$$

There is no need to put ϕ twice on the left of $\vdash$, since Γ in a sequent $(\Gamma \vdash \psi)$ is a set, not a sequence.

Remark 2.2.5 The derivation (2.9) is a *proof of* its conclusion $((\phi \wedge \psi) \wedge \chi)$ from certain assumptions. It is also a *proof of* the sequent (2.10), by showing that (2.10) is correct. In mathematics this is par for the course; the same argument can be used to establish many different things. But in logic, where we are comparing different proofs all the time, there is a danger of confusion. For mental hygiene we shall say that (2.9) is a *derivation of* its conclusion, but a *proof of* the sequent (2.10).

Exercises

2.2.1. Express the following using $\wedge$ between statements:

   (a) The real number r is positive but not an integer.

   (b) v is a nonzero vector.

   (c) ϕ if and only if ψ. [Here ϕ and ψ stand for statements.]

2.2.2. Write out derivations that prove the following sequents:

   (a) $\{\phi, \psi, \chi\} \vdash (\phi \wedge (\psi \wedge \chi))$.

   (b) $\{\phi, \psi\} \vdash (\psi \wedge \phi)$.

   (c) $\{\phi\} \vdash ((\phi \wedge \phi) \wedge \phi)$.

   (d) $\{\phi, \psi\} \vdash ((\phi \wedge \psi) \wedge (\phi \wedge \psi))$.

2.3 Arguments eliminating 'and'

Often in arguments we rely on a statement of the form (ϕ and ψ) to justify the next step in the argument. The simplest examples are where the next step is to deduce ϕ, or to deduce ψ.

Example 2.3.1 We prove that every prime greater than 2 is odd. *Let p be a prime greater than 2.* Since *p is prime*, p is not divisible by any integer n with $1 < n < p$. Since p is greater than 2, $1 < 2 < p$. So p is not divisible by 2, in other words, p is odd.

In this argument we assume

(2.12)   (p is prime $\wedge$ p is greater than 2)

(the first passage in italics). From (2.12) we deduce

(2.13) p is prime

(the second passage in italics).

Reflecting on this example, we extract another natural deduction rule:

NATURAL DEDUCTION RULE ($\wedge$E)   If

$$D$$
$$(\phi \wedge \psi)$$

is a derivation of ($\phi \wedge \psi$), then

$$\frac{\begin{matrix} D \\ (\phi \wedge \psi) \end{matrix}}{\phi} \text{ ($\wedge$E)} \quad \text{and} \quad \frac{\begin{matrix} D \\ (\phi \wedge \psi) \end{matrix}}{\psi} \text{ ($\wedge$E)}$$

are derivations of ϕ and ψ, respectively. Their undischarged assumptions are those of D.

In the label ($\wedge$E) the E stands for Elimination, and this rule is known as $\wedge$-elimination. The reason is that in the derivations of ϕ and ψ, the occurrence of the symbol $\wedge$ in the middle of ($\phi \wedge \psi$) is eliminated in the conclusion (together with one of the statements ϕ and ψ). This is the opposite to ($\wedge$I), where an occurrence of $\wedge$ is introduced in the conclusion.

In sequent terms, this natural deduction rule tells us:

SEQUENT RULE ($\wedge$E)   If the sequent ($\Gamma \vdash (\phi \wedge \psi)$) is correct, then so are both the sequents ($\Gamma \vdash \phi$) and ($\Gamma \vdash \psi$).

We can use both of the rules ($\wedge$I) and ($\wedge$E) in a single derivation, for example:

Example 2.3.2

$$\dfrac{\dfrac{(\phi \wedge \psi)}{\psi}\ (\wedge\mathrm{E}) \qquad \dfrac{(\phi \wedge \psi)}{\phi}\ (\wedge\mathrm{E})}{(\psi \wedge \phi)}\ (\wedge\mathrm{I})$$

This derivation proves the sequent $\{(\phi \wedge \psi)\} \vdash (\psi \wedge \phi)$.

Example 2.3.3

$$\dfrac{\dfrac{(\phi \wedge (\psi \wedge \chi))}{\phi}\ (\wedge\mathrm{E}) \qquad \dfrac{\dfrac{(\phi \wedge (\psi \wedge \chi))}{(\psi \wedge \chi)}\ (\wedge\mathrm{E})}{\psi}\ (\wedge\mathrm{E})}{(\phi \wedge \psi)}\ (\wedge\mathrm{I}) \qquad \dfrac{\dfrac{(\phi \wedge (\psi \wedge \chi))}{(\psi \wedge \chi)}\ (\wedge\mathrm{E})}{\chi}\ (\wedge\mathrm{E})}{((\phi \wedge \psi) \wedge \chi)}\ (\wedge\mathrm{I})$$

This derivation proves the sequent $\{(\phi \wedge (\psi \wedge \chi))\} \vdash ((\phi \wedge \psi) \wedge \chi)$.

Exercises

2.3.1. Write out derivations that prove the following sequents:
  (a) $\{(\phi \wedge \psi)\} \vdash (\phi \wedge \phi)$.
  (b) $\{((\phi \wedge \psi) \wedge \chi)\} \vdash (\phi \wedge (\psi \wedge \chi))$.
  (c) $\{\phi, (\psi \wedge \chi)\} \vdash (\chi \wedge \phi)$.
  (d) $\{(\phi \wedge (\psi \wedge \chi))\} \vdash ((\chi \wedge \phi) \wedge \psi)$.

2.3.2. Fill in the blanks (marked $\star$) in the derivation below. Then show that the derivation can be shortened to a derivation with the same conclusion and no extra assumptions, but with fewer applications of ($\wedge$I) and ($\wedge$E).

$$\dfrac{\dfrac{\begin{array}{c}D\\\star\end{array} \qquad \begin{array}{c}D'\\\psi\end{array}}{\star}\ (\wedge\mathrm{I})}{\phi}\ (\wedge\mathrm{E})$$

[The moral is that there is never any point in doing an ($\wedge$E) immediately after an ($\wedge$I).]

2.3.3. Show that $\{\phi_1, \phi_2\} \vdash \psi$ if and only if $\{(\phi_1 \wedge \phi_2)\} \vdash \psi$.

2.4 Arguments using 'if'

We write $(\phi \rightarrow \psi)$ for 'If ϕ then ψ', where ϕ and ψ are statements. There are two natural deduction rules for $\rightarrow$, an introduction rule ($\rightarrow$I) and an elimination rule ($\rightarrow$E). As with $\wedge$, these rules are based on the ways that we use the words 'if...then' in arguments.

We begin with the introduction rule. How does one prove a conclusion of the form 'If ϕ then ψ'? Here is a typical example.

Example 2.4.1 Write p for the statement that if x is real then $x^2 + 1 \geqslant 2x$. We prove p as follows. Assume x is real. Then $x - 1$ is real, so

$$0 \leqslant (x-1)^2 = x^2 - 2x + 1 = (x^2 + 1) - 2x$$

So

$$2x \leqslant x^2 + 1 \quad \square$$

It may help to arrange this proof in a diagram:

(2.14)

> Assume x is real.
>
> Then $x - 1$ is real, so
>
> $0 \leqslant (x-1)^2 = x^2 - 2x + 1 = (x^2 + 1) - 2x$
>
> So, $2x \leqslant x^2 + 1$.

So, if x is real then $x^2 + 1 \geqslant 2x$.

We have two proofs here. The larger proof consists of the whole of (2.14), and its conclusion is

(2.15)   If x is real then $x^2 + 1 \geqslant 2x$.

The smaller proof is inside the box, and its conclusion is

(2.16) $2x \leqslant x^2 + 1$.

The smaller proof assumes that x is real. But the larger proof does not assume anything about x. Even if x is $\sqrt{-1}$, it is still true (but uninteresting) that *if x is real then $x^2 + 1 \geqslant 2x$*. We say that in the larger proof the assumption 'x is real' is *discharged*—it is no longer needed, because the assumption has been put as the 'if' part of the conclusion.

In the natural deduction calculus we have a notation for discharging assumptions. Every assumption of a derivation D is written somewhere in D, perhaps in several places. (Remember from Section 2.1 that for logicians it is important to make our assumptions explicit.) We shall discharge *occurrences* of assumptions. We do it by writing a line through the text of the assumption. We call this line a *dandah*. (This is the authors' terminology, taken from Sanskrit grammar; there is no standard name for the discharging symbol.)

Thus if ϕ is an assumption written somewhere in D, then we discharge ϕ by writing a dandah through it: $\phi\!\!\!/$. In the rule ($\rightarrow$I) below, and in similar rules later, the $\phi\!\!\!/$ means that in forming the derivation *we are allowed to discharge any occurrences of the assumption ϕ written in D*. The rule is still correctly applied if we do not discharge all of them; in fact the rule is correctly applied even if ϕ is not an assumption of D at all, so that there is nothing to discharge. Example 2.4.4 will illustrate these points.

NATURAL DEDUCTION RULE ($\rightarrow$I) Suppose

$$D$$
$$\psi$$

is a derivation of ψ, and ϕ is a statement. Then the following is a derivation of $(\phi \rightarrow \psi)$:

$$\phi\!\!\!/$$
$$D$$
$$\frac{\psi}{(\phi \rightarrow \psi)} \quad (\rightarrow\text{I})$$

Its undischarged assumptions are those of D, except possibly ϕ.

We can also express ($\rightarrow$I) as a sequent rule:

SEQUENT RULE ($\rightarrow$I) If the sequent $(\Gamma \cup \{\phi\} \vdash \psi)$ is correct then so is the sequent $(\Gamma \vdash (\phi \rightarrow \psi))$.

Discharging is a thing that happens in derivations, not in sequents. Instead of being discharged, the assumption ϕ in the first sequent of the sequent rule ($\rightarrow$I) is allowed to drop out of the assumptions of the second sequent. But note that ϕ could be one of the assumptions in Γ, and in this case it will still be an assumption of the second sequent. This corresponds to the fact that the natural deduction rule ($\rightarrow$I) *allows* us to discharge any occurrence of the assumption ϕ when we form the derivation of $(\phi \rightarrow \psi)$.

Remark 2.4.2 Thanks to the rule ($\rightarrow$I), a derivation D can have assumptions of two kinds: discharged assumptions and undischarged assumptions. (It is quite possible for the same statement to appear as a discharged assumption in one

Antoine Arnauld France, 1612–1694.
His *Port-Royal Logic*, written in 1662 with Pierre
Nicole, introduced the rule ($\rightarrow$I) as a way of
rewriting arguments to make them more 'beautiful'.

part of a derivation and an undischarged assumption in another.) We say that
D is a *derivation of* its conclusion *from* its undischarged assumptions. We count
D as proving a sequent $\Gamma \vdash \psi$ when ψ is the conclusion of D and all the *undischarged* assumptions of D lie in Γ. The discharged assumptions have fulfilled
their purpose by being discharged when ($\rightarrow$I) is used, and they need not be
mentioned again. In particular, we can now have derivations with no undischarged assumptions at all; these derivations prove sequents of the form ($\vdash \phi$)
(recall Definition 2.1.2).

We can combine the rule ($\rightarrow$I) with the rules for $\wedge$ to give new derivations.

Example 2.4.3 A proof of the sequent $\vdash (\phi \rightarrow (\psi \rightarrow (\phi \wedge \psi)))$:

$$
\begin{array}{c}
② \qquad\qquad ① \\
\dfrac{\phi \qquad\qquad \psi}{\begin{array}{c} ① \dfrac{(\phi \wedge \psi)}{② \dfrac{(\psi \rightarrow (\phi \wedge \psi))}{(\phi \rightarrow (\psi \rightarrow (\phi \wedge \psi)))}} \ (\rightarrow\text{I}) \end{array}} \ (\wedge\text{I})
\end{array}
$$

In this derivation the top step is an application of ($\wedge$I) to assumptions ϕ
and ψ. Since ϕ and ψ are assumed, we have no proofs to write above them, so
we do not write the D and D' of the rule ($\wedge$I). Next we extend the derivation
downwards by applying ($\rightarrow$I) once, adding '$\psi \rightarrow$' on the left of the conclusion.
This allows us to discharge the assumption ψ. To show that ψ is discharged at
this step, we number the step 1 (by writing 1 in a circle at the left side of the
step), and we attach the same label '1' to the dandah through ψ at top right.
Finally we apply ($\rightarrow$I) once more; this discharges ϕ, and so we number the step
2 and put the label 2 on the dandah through ϕ at top left.

Example 2.4.4 Here we prove the sequent $\vdash (\phi \to (\phi \to \phi))$.

$$
\underset{\textcircled{2}}{} \underset{(\phi \to (\phi \to \phi))}{\dfrac{\underset{\textcircled{1}}{} \dfrac{\overset{\overset{?}{\phi}}{}}{(\phi \to \phi)} \; (\to\text{I})}{}} \; (\to\text{I})
$$

The sequent has no assumptions, so we need to discharge ϕ somewhere. But there are two steps where we can discharge it, and it does not matter which we use. We could discharge it when we first apply ($\to$I), so that '?' becomes 1; in this case there is nothing to discharge at the second application of ($\to$I). Alternatively we could leave it undischarged at the first application of ($\to$I) and discharge it at the second, writing 2 for '?'. Both ways are correct.

Turning to the rule for eliminating 'if', suppose we have proved something of the form 'If ϕ then ψ'. How do we use this to deduce something else? Suppose, for example, that we have proved a lemma saying

(2.17)   If $q \leqslant 1$ then $f(q) = \pi$

The most straightforward way to apply (2.17) is to prove $q \leqslant 1$, and then deduce from (2.17) that $f(q) = \pi$. In short, if we have proved both ϕ and 'If ϕ then ψ', then we can deduce ψ. This idea goes over straightforwardly into the following natural deduction rule.

NATURAL DEDUCTION RULE ($\to$E)   If

$$
\begin{array}{ccc} D & & D' \\ \phi & \text{and} & (\phi \to \psi) \end{array}
$$

are derivations of ϕ and $(\phi \to \psi)$, respectively, then

$$
\dfrac{\begin{array}{cc} D & D' \\ \phi & (\phi \to \psi) \end{array}}{\psi} \; (\to\text{E})
$$

is a derivation of ψ. Its undischarged assumptions are those of D together with those of D'.

In sequent terms:

SEQUENT RULE ($\to$E)   If $(\Gamma \vdash \phi)$ and $(\Delta \vdash (\phi \to \psi))$ are both correct sequents, then the sequent $(\Gamma \cup \Delta \vdash \psi)$ is correct.

We can combine ($\to$E) with other rules to make various derivations.

Example 2.4.5 A derivation to prove the sequent $\{(\phi \to \psi), (\psi \to \chi)\} \vdash (\phi \to \chi)$.

$$\cfrac{\cfrac{\not\phi^{①} \qquad (\phi \to \psi)}{\psi} \,(\to E) \qquad (\psi \to \chi)}{\cfrac{\chi}{① \; (\phi \to \chi)} \,(\to I)} \,(\to E)$$

Exercises

2.4.1. Write the following using $\to$ between statements:

(a) f is continuous if f is differentiable.

(b) Supposing x is positive, x has a square root.

(c) $ab/b = a$ provided $a \neq 0$.

2.4.2. In the following two derivations, the names of the rules are missing, and so are the dandahs and step numbers for the assumptions that are discharged. Write out the derivations, including these missing pieces.

(a) A proof of $\vdash ((\phi \wedge \psi) \to (\psi \wedge \phi))$

$$\cfrac{\cfrac{(\phi \wedge \psi)}{\psi} \qquad \cfrac{(\phi \wedge \psi)}{\phi}}{\cfrac{(\psi \wedge \phi)}{((\phi \wedge \psi) \to (\psi \wedge \phi))}}$$

(b) A proof of $\vdash ((\psi \to \chi) \to ((\phi \to \psi) \to (\phi \to \chi)))$

$$\cfrac{\cfrac{\cfrac{\cfrac{\phi \qquad (\phi \to \psi)}{\psi} \qquad (\psi \to \chi)}{\chi}}{\cfrac{(\phi \to \chi)}{((\phi \to \psi) \to (\phi \to \chi))}}}{((\psi \to \chi) \to ((\phi \to \psi) \to (\phi \to \chi)))}$$

2.4.3. Each of the following derivations proves a sequent. Write out the sequent that it proves.

(a)

$$\dfrac{\dfrac{\dfrac{\phi^{①}}{(\psi \to \phi)}\ (\to I)}{(\phi \to (\psi \to \phi))}}{}\ (\to I).$$
①

(b)

$$\dfrac{\dfrac{\dfrac{\phi}{(\psi \to \phi)}\ (\to I)}{(\phi \to (\psi \to \phi))}}{}\ (\to I).$$

(c)

$$\dfrac{\dfrac{\dfrac{\psi^{①} \qquad \dfrac{(\phi \wedge \psi)}{\phi}\ (\wedge E)}{(\psi \wedge \phi)}\ (\wedge I)}{(\psi \to (\psi \wedge \phi))}\ (\to I)}{}$$
① ①

(d)

$$\dfrac{\dfrac{\phi^{①}}{(\phi \to \phi)}}{}\ (\to I).$$
①

2.4.4. Write out derivations to prove each of the following sequents.

(a) $\vdash (\phi \to (\psi \to \psi))$.

(b) $\vdash ((\phi \to \phi) \wedge (\psi \to \psi))$.

(c) $\vdash ((\phi \to (\theta \to \psi)) \to (\theta \to (\phi \to \psi)))$.

(d) $\{(\phi \to \psi), (\phi \to \chi)\} \vdash (\phi \to (\psi \wedge \chi))$.

(e) $\{(\phi \to \psi), ((\phi \wedge \psi) \to \chi)\} \vdash (\phi \to \chi)$.

(f) $\{(\phi \to (\psi \to \chi))\} \vdash ((\phi \wedge \psi) \to \chi)$.

(g) $\vdash ((\phi \to \psi) \to ((\psi \to \theta) \to (\phi \to \theta)))$.

(h) $\vdash ((\phi \to (\psi \wedge \theta)) \to ((\phi \to \theta) \wedge (\phi \to \psi)))$.

2.4.5 Show that $\{\phi\} \vdash \psi$ if and only if $\vdash (\phi \to \psi)$. [Prove the directions $\Rightarrow$ and $\Leftarrow$ separately.]

2.4.6 Let ϕ and ψ be statements and Γ a set of statements. Consider the two sequents

(a) $\Gamma \cup \{\phi\} \vdash \psi$.

(b) $\Gamma \vdash (\phi \to \psi)$.

Show that if D_1 is a derivation proving (a), then D_1 can be used to construct a derivation D_1' proving (b). Show also that if D_2 is a derivation proving (b), then D_2 can be used to construct a derivation D_2' proving (a). (Together these show that (a) has a proof by derivation if and only if (b) has a proof by derivation. The previous exercise is the special case where Γ is empty.)

2.5 Arguments using 'if and only if'

We shall write

(2.18) $$(\phi \leftrightarrow \psi)$$

for 'ϕ if and only if ψ'. We saw already in Exercise 2.2.1(c) that 'ϕ if and only if ψ' expresses the same as

(2.19) $$(\text{if } \phi \text{ then } \psi \wedge \text{if } \psi \text{ then } \phi)$$

Thanks to this paraphrase, we can use the introduction and elimination rules for $\wedge$ to devise introduction and elimination rules for $\leftrightarrow$, as follows.

NATURAL DEDUCTION RULE ($\leftrightarrow$I) If

$$\begin{array}{ccc} D & & D' \\ (\phi \rightarrow \psi) & \text{and} & (\psi \rightarrow \phi) \end{array}$$

are derivations of $(\phi \rightarrow \psi)$ and $(\psi \rightarrow \phi)$, respectively, then

$$\frac{\begin{array}{cc} D & D' \\ (\phi \rightarrow \psi) & (\psi \rightarrow \phi) \end{array}}{(\phi \leftrightarrow \psi)} \; (\leftrightarrow\text{I})$$

is a derivation of $(\phi \leftrightarrow \psi)$. Its undischarged assumptions are those of D together with those of D'.

NATURAL DEDUCTION RULE ($\leftrightarrow$E) If

$$\begin{array}{c} D \\ (\phi \leftrightarrow \psi) \end{array}$$

is a derivation of $(\phi \leftrightarrow \psi)$, then

$$\frac{\begin{array}{c} D \\ (\phi \leftrightarrow \psi) \end{array}}{(\phi \rightarrow \psi)} \; (\leftrightarrow\text{E}) \quad \text{and} \quad \frac{\begin{array}{c} D' \\ (\phi \leftrightarrow \psi) \end{array}}{(\psi \rightarrow \phi)} \; (\leftrightarrow\text{E})$$

are derivations of $(\phi \rightarrow \psi)$ and $(\psi \rightarrow \phi)$, respectively. Their undischarged assumptions are those of D.

Example 2.5.1 A proof of the sequent $\{(\phi \leftrightarrow \psi)\} \vdash (\psi \leftrightarrow \phi)$. (Compare Example 2.3.2, and make sure to get the left and right branches of the derivation the right way round.)

$$\frac{\dfrac{(\phi \leftrightarrow \psi)}{(\psi \rightarrow \phi)} \ (\leftrightarrow\text{E}) \qquad \dfrac{(\phi \leftrightarrow \psi)}{(\phi \rightarrow \psi)} \ (\leftrightarrow\text{E})}{(\psi \leftrightarrow \phi)} \ (\leftrightarrow\text{I})$$

Exercises

2.5.1.   Give derivations to prove the following sequents:

(a) $\{\phi, (\phi \leftrightarrow \psi)\} \vdash \psi$.

(b) $\vdash (\phi \leftrightarrow \phi)$.

(c) $\{(\phi \leftrightarrow \psi), (\psi \leftrightarrow \chi)\} \vdash (\phi \leftrightarrow \chi)$.

(d) $\vdash ((\phi \leftrightarrow (\psi \leftrightarrow \chi)) \leftrightarrow ((\phi \leftrightarrow \psi) \leftrightarrow \chi))$.

(e) $\{(\phi \leftrightarrow (\psi \leftrightarrow \psi))\} \vdash \phi$.

2.5.2.   Let S be any set of statements, and let $\sim$ be the relation on S defined by: for all $\phi, \psi \in S$,

$$\phi \sim \psi \quad \text{if and only if} \quad \vdash (\phi \leftrightarrow \psi).$$

Show that $\sim$ is an equivalence relation on S. That is, it has the three properties:

- (Reflexive) For all ϕ in S, $\phi \sim \phi$.
- (Symmetric) For all ϕ and ψ in S, if $\phi \sim \psi$ then $\psi \sim \phi$.
- (Transitive) For all ϕ, ψ and χ in X, if $\phi \sim \psi$ and $\psi \sim \chi$ then $\phi \sim \chi$.

[For reflexivity use (b) of Exercise 2.5.1. With a little more work, (c) and Example 2.5.1 give transitivity and symmetry.]

2.5.3   Show that if we have a derivation D of ψ with no undischarged assumptions, then we can use it to construct, for any statement ϕ, a derivation of $((\phi \leftrightarrow \psi) \leftrightarrow \phi)$ with no undischarged assumptions.

2.5.4   Devise suitable sequent rules for $\leftrightarrow$.

2.6 Arguments using 'not'

The rules for 'not' and 'or' are not quite as straightforward as the rules we have encountered so far. But in the spirit of natural deduction, they do all correspond to moves commonly made in mathematical arguments. We consider 'not' in this section and 'or' in the next.

If ϕ is a statement, we write $(\neg\phi)$ for the statement expressing that ϕ is not true. The symbol $\neg$ is pronounced 'not' or 'negation', and $(\neg\phi)$ is called the *negation* of ϕ.

How is $\neg$ used in arguments? We take our first cue not from the mathematicians, but from a statement of Ian Hislop, the editor of the journal *Private Eye*. In 1989 the journal lost a libel case and was instructed to pay £600,000 damages. Coming out of the trial, Ian Hislop stood on the courthouse steps and said

(2.20)   If that's justice then I'm a banana.

He meant 'That's not justice'. He was using the following device.

We write $\perp$ (pronounced 'absurdity' or 'bottom' according to taste) for a statement which is definitely false, for example, '$0 = 1$' or 'I'm a banana'. In derivations we shall treat $(\neg\phi)$ exactly as if it was written $(\phi \to \perp)$.

How does this work in practice? Suppose first that we have proved or assumed $(\neg\phi)$. Then we can proceed as if we proved or assumed $(\phi \to \perp)$. The rule $(\to E)$ tells us that from ϕ and $(\phi \to \perp)$ we can deduce $\perp$. So we will deduce $\perp$ from ϕ and $(\neg\phi)$.

This gives us our first natural deduction rule for $\neg$:

NATURAL DEDUCTION RULE $(\neg E)$ If

$$
\begin{array}{ccc}
D & & D' \\
\phi & \text{and} & (\neg\phi)
\end{array}
$$

are derivations of ϕ and $(\neg\phi)$ respectively, then

$$
\dfrac{\begin{array}{cc} D & D' \\ \phi & (\neg\phi) \end{array}}{\perp} \; (\neg E)
$$

is a derivation of $\perp$. Its undischarged assumptions are those of D together with those of D'.

Second, suppose we want to prove $(\neg\phi)$. Then we proceed as if we were using $(\to I)$ to prove $(\phi \to \perp)$. In other words, we assume ϕ and deduce $\perp$. The assumption ϕ is discharged after it has been used.

NATURAL DEDUCTION RULE (¬I)   Suppose

$$
\begin{array}{c}
D \\
\bot
\end{array}
$$

is a derivation of $\bot$, and ϕ is a statement. Then the following is a derivation of $(\neg\phi)$:

$$
\begin{array}{c}
\cancel{\phi} \\
D \\
\bot \\
\hline
(\neg\phi)
\end{array} \text{(¬I)}
$$

Its undischarged assumptions are those of D, except possibly ϕ.

Example 2.6.1 The following derivation proves $\vdash (\phi \to (\neg(\neg\phi)))$.

$$
\cfrac{\cancel{\phi}^{②} \qquad \cancel{(\neg\phi)}^{①}}{\cfrac{\bot}{(\neg(\neg\phi))} \text{(¬I)} \;①}{\cfrac{}{(\phi \to (\neg(\neg\phi)))} \text{(→I)} \;②}} \text{(¬E)}
$$

At the application of (¬I) we discharge the assumption $(\neg\phi)$ to get the conclusion $(\neg(\neg\phi))$.

Just to say that $(\neg\phi)$ behaves like $(\phi \to \bot)$, without adding anything about how $\bot$ behaves, leaves rather a lot unexplained. Surprisingly, we were able to carry out the derivation in Example 2.6.1 without using any information about $\bot$. But to go any further, we need to know how absurdity behaves in proofs. The following argument is a case in point.

Example 2.6.2

Theorem *There are infinitely many prime numbers.*

Proof Assume not. Then there are only finitely many prime numbers

$$
p_1, \ldots, p_n
$$

Consider the integer

$$
q = (p_1 \times \cdots \times p_n) + 1
$$

The integer q must have at least one prime factor r. But then r is one of the p_i, so it cannot be a factor of q. Hence r both is and is not a factor of q; absurd! So our assumption is false, and the theorem is true. $\square$

A close inspection of this argument shows that we prove the theorem ϕ by assuming $(\neg\phi)$ and deducing an absurdity. The assumption $(\neg\phi)$ is then discharged. This form of argument is known as *reductio ad absurdum*, RAA for short. In natural deduction terms it comes out as follows.

NATURAL DEDUCTION RULE (RAA)   Suppose we have a derivation

$$
\begin{array}{c}
D \\
\bot
\end{array}
$$

whose conclusion is $\bot$. Then there is a derivation

$$
\begin{array}{c}
\cancel{(\neg\phi)} \\
D \\
\dfrac{\bot}{\phi} \ \text{(RAA)}
\end{array}
$$

Its undischarged assumptions are those of D, except possibly $(\neg\phi)$.

Example 2.6.3 We prove $\vdash ((\neg(\neg\phi)) \to \phi)$.

$$
\cfrac{\cfrac{\cancel{(\neg\phi)}^{\ \textcircled{1}} \qquad\qquad \cancel{(\neg(\neg\phi))}^{\ \textcircled{2}}}{\textcircled{1}\ \dfrac{\bot}{\phi}\ \text{(RAA)}}{\ \textcircled{2}\ \ \dfrac{}{((\neg(\neg\phi)) \to \phi)}} \ \ \substack{(\neg E)}
$$

Although Examples 2.6.2 and 2.6.3 are reasonably straightforward, the use of (RAA) can lead to some very unintuitive proofs. Generally, it is the rule of last resort, if you cannot find anything else that works.

Here are the sequent rules corresponding to $(\neg E)$, $(\neg I)$ and (RAA):

SEQUENT RULE $(\neg E)$ If $(\Gamma \vdash \phi)$ and $(\Delta \vdash (\neg\phi))$ are both correct sequents, then the sequent $(\Gamma \cup \Delta \vdash \bot)$ is correct.

SEQUENT RULE $(\neg I)$ If the sequent $(\Gamma \cup \{\phi\} \vdash \bot)$ is correct, then so is the sequent $(\Gamma \vdash (\neg\phi))$.

SEQUENT RULE (RAA) If the sequent $(\Gamma \cup \{(\neg\phi)\} \vdash \bot)$ is correct, then so is the sequent $(\Gamma \vdash \phi)$.

Exercises

2.6.1. Find natural deduction proofs for the following sequents (none of which need (RAA)):

(a) $\vdash (\neg(\phi \wedge (\neg\phi)))$.

(b) $\vdash ((\neg(\phi \rightarrow \psi)) \rightarrow (\neg\psi))$.

(c) $\vdash ((\phi \wedge \psi) \rightarrow (\neg(\phi \rightarrow (\neg\psi))))$.

(d) $\{((\neg(\phi \wedge \psi)) \wedge \phi)\} \vdash (\neg\psi)$.

(e) $\{(\phi \rightarrow \psi)\} \vdash ((\neg\psi) \rightarrow (\neg\phi))$.

(f) $\{(\phi \rightarrow \psi)\} \vdash (\neg(\phi \wedge (\neg\psi)))$.

2.6.2. Find natural deduction proofs for the following sequents (all of which need (RAA)):

(a) $\{((\neg\psi) \rightarrow (\neg\phi))\} \vdash (\phi \rightarrow \psi)$.
    [Assume ϕ and $((\neg\psi) \rightarrow (\neg\phi))$. Prove ψ by (RAA), assuming $(\neg\psi)$ and deducing $\bot$.]

(b) $\vdash ((\neg(\phi \rightarrow \psi)) \rightarrow \phi)$.

(c) $\vdash (\phi \rightarrow ((\neg\phi) \rightarrow \psi))$.

(d) $\{(\neg(\phi \leftrightarrow \psi))\} \vdash ((\neg\phi) \leftrightarrow \psi)$. (Hard.)

2.7 Arguments using 'or'

We write $(\phi \vee \psi)$ for 'Either ϕ or ψ or both'. The symbol '$\vee$' is read as 'or'. For example, $(x = 0 \vee x > 0)$ says

Either x is 0 or x is greater than 0 or both.

In this case (as often) the 'or both' doesn't arise and can be ignored. The whole statement is equivalent to '$x \geqslant 0$'.

There are introduction and elimination rules for $\vee$. The introduction rule is much easier to handle than the elimination rule.

SEQUENT RULE ($\vee$I) If at least one of $(\Gamma \vdash \phi)$ and $(\Gamma \vdash \psi)$ is a correct sequent, then the sequent $(\Gamma \vdash (\phi \vee \psi))$ is correct.

NATURAL DEDUCTION RULE ($\vee$I) If

$$D$$
$$\phi$$

is a derivation with conclusion ϕ, then

$$\frac{\begin{array}{c} D \\ \phi \end{array}}{(\phi \vee \psi)}$$

is a derivation of $(\phi \vee \psi)$. Its undischarged assumptions are those of D. Similarly if

$$\begin{array}{c} D \\ \psi \end{array}$$

is a derivation with conclusion ψ, then

$$\frac{\begin{array}{c} D \\ \psi \end{array}}{(\phi \vee \psi)}$$

is a derivation with conclusion $(\phi \vee \psi)$. Its undischarged assumptions are those of D.

Example 2.7.1 We prove the sequent $\vdash (\neg(\neg(\phi \vee (\neg\phi))))$. Since the conclusion begins with $\neg$, common sense suggests we try using $(\neg I)$ to derive it. In other words, we should try first to derive $\bot$ from $(\neg(\phi \vee (\neg\phi)))$. We can do this by first proving $(\phi \vee (\neg\phi))$ from the assumption $(\neg(\phi \vee (\neg\phi)))$. This looks like a curious thing to do, but it works:

(2.21)

At first sight it looks as if the two $\neg$ signs at the beginning of the conclusion have made extra work for us. This is not so. The sequent $(\vdash (\phi \vee (\neg\phi)))$ is certainly valid, but it is just as hard to prove; in fact the proof needs (RAA).

You might guess that the sequent $(\vdash (\phi \vee (\neg\phi)))$ could be proved immediately by $(\vee I)$. But reflection shows that this would involve proving either ϕ or $(\neg\phi)$, and since we have not said what statement ϕ is, it is hardly likely that we could prove either one of ϕ and $(\neg\phi)$. This leaves us with no obvious strategy. In such cases, a sensible move is to try to prove a contradiction from the negation of the conclusion, and then finish with (RAA). The derivation below does exactly this.

$$\begin{array}{c}
\cfrac{(\neg\phi)^{\;①}}{(\phi\vee(\neg\phi))}\;(\vee\mathrm{I}) \qquad (\neg(\phi\vee(\neg\phi)))^{\;③} \\
\hline
①\;\cfrac{\bot}{\phi}\;(\mathrm{RAA})
\end{array}\;(\neg\mathrm{E})$$

(2.22)

We turn to the elimination rule for $\vee$. How do we use assumptions of the form $(\phi\vee\psi)$ in mathematical arguments? Here is an example. We leave out the technical details—they can be found in calculus texts.

Example 2.7.2 Consider how we show that if n is an integer $\neq 0$ and $x \neq 0$ then

$$\frac{dx^n}{dx} = nx^{n-1}.$$

There is a well-known proof when n is positive. But this argument will not work when n is negative. So a different argument is needed for this case. The resulting proof has the following form

> We assume $n \neq 0$, and so either $n > 0$ or $n < 0$.

> Case One: $n > 0$, etc., so $dx^n/dx = nx^{n-1}$.
> Case Two: $n < 0$, etc., so $dx^n/dx = nx^{n-1}$.

> Since the conclusion holds in both cases, and at least one of the cases must apply, the conclusion holds.

NATURAL DEDUCTION RULE ($\vee$E)   Given derivations

$$\begin{array}{ccc}
D & D' & D'' \\
(\phi\vee\psi)\;, & \chi & \text{and} \quad \chi
\end{array}$$

we have a derivation

$$\cfrac{\begin{array}{ccc}
& \overset{\phi}{} & \overset{\psi}{} \\
D & D' & D'' \\
(\phi\vee\psi) & \chi & \chi
\end{array}}{\chi}$$

Its undischarged assumptions are those of D, those of D' except possibly ϕ, and those of D'' except possibly ψ.

SEQUENT RULE ($\vee$E)   If $(\Gamma \cup \{\phi\} \vdash \chi)$ and $(\Delta \cup \{\psi\} \vdash \chi)$ are correct sequents, then the sequent $(\Gamma \cup \Delta \cup \{(\phi \vee \psi)\} \vdash \chi)$ is correct.

Exercises

2.7.1. Give natural deduction proofs of the following sequents (none of which need ($\vee$E)):

(a) $\vdash (\phi \to (\phi \vee \psi))$.

(b) $\{(\neg(\phi \vee \psi))\} \vdash ((\neg\phi) \wedge (\neg\psi))$.

(c) $\vdash ((\phi \to \psi) \to ((\neg\phi) \vee \psi))$.

2.7.2. Give natural deduction proofs of the following sequents. (These need ($\vee$E).)

(a) $\{(\phi \vee \psi)\} \vdash (\psi \vee \phi)$.

(b) $\{(\phi \vee \psi), (\phi \to \chi), (\psi \to \chi)\} \vdash \chi$.

(c) $\{(\phi \vee \psi), (\neg\phi)\} \vdash \psi$.

(d) $\{((\neg\phi) \wedge (\neg\psi))\} \vdash (\neg(\phi \vee \psi))$.

(e) $\{(\phi \wedge \psi)\} \vdash (\neg((\neg\phi) \vee (\neg\psi)))$.

3 Propositional logic

We have now proved a wide range of statements. For example, in Section 2.4 we proved the correctness of the sequent $\vdash (\phi \to (\psi \to (\phi \wedge \psi)))$ (Example 2.4.3).

Strictly this is not correct. The Greek letters 'ϕ' and 'ψ' are not statements; they are *variables ranging over statements* (just as x, y can be variables ranging over real numbers). We could put any statement in place of ϕ and any statement in place of ψ.

Nevertheless, we did prove something. What we proved in Example 2.4.3 was that

Every statement of the form $(\phi \to (\psi \to (\phi \wedge \psi)))$ is true.

Likewise the natural deduction 'proof' of $(\phi \to (\psi \to \phi))$ is strictly not a proof of a statement; it is a pattern of infinitely many proofs of different statements. Such patterns are called *formal proofs*.

So we are studying *patterns* that infinitely many different statements could have. These patterns are the real subject matter of mathematical logic. To study them closer, we work with *formal languages* which are designed to express the patterns that are important for arguments.

To design a language, we need to describe three things.

- The lexicon is the set of 'words' of the language. In a formal language the words are called *symbols*.

- The syntax describes how the words are built up into sentences. A sentence of a formal language is called a *formula*. (In Chapter 5 we will introduce a particular class of formulas that we call 'sentences'. Having two meanings of 'sentence' is unfortunate, but the terminology is well established and there is little danger of confusion.)

- The semantics is the correlation between symbols and their meanings. In our formal languages the symbols fall into two classes. Some of the symbols are symbolisations of expressions in ordinary English, such as 'and', 'not' and 'for all'. These symbols could be—and some of them often are—used in ordinary mathematical writing. We can study their meanings by seeing how they are used in mathematical arguments. These are the 'pattern' words. Other symbols of our language have no fixed meaning, but we have ways of attaching temporary meanings to them. In the formal semantics of logical languages, we give formal necessary and sufficient conditions for a statement to be true, depending on how the temporary meanings are assigned.

Augustus De Morgan London, 1806–1871.
He proposed the name 'mathematical logic' for the mathematical study of patterns that guarantee the correctness of arguments.

You should bear in mind throughout this chapter that we will be doing *calculations*. The meanings of symbols may motivate this or that calculation, but the calculations themselves do not involve meanings; they operate entirely with the lexicon and the syntax.

3.1 LP, the language of propositions

We begin with LP, a formal language for expressing propositions.

The first step in constructing LP is to choose a set of symbols that can stand for statements. We call this set the *signature* (or more strictly the *propositional signature*, though in this chapter we use the shorter name). In theory, it can be any set of symbols; but in practice we should avoid putting into it symbols that already have other uses. We should certainly not put into it any of the symbols

(3.1) $\land \quad \lor \quad \rightarrow \quad \leftrightarrow \quad \neg \quad \bot$

(we call these the *truth function* symbols) and the parentheses

(3.2) $(\quad)$

The symbols in the signature are called *propositional symbols*. The usual custom is to choose them to be lower-case letters near p, sometimes with indices or dashes, for example,

(3.3) $p \quad q \quad r \quad p_1 \quad q_{215} \quad r' \quad r''''$

In computer science applications of logic one often meets propositional symbols that are several letters long, like MouseEvent. The infinite set of symbols

(3.4) $p_0, p_1, p_2, \ldots$

will serve as a default signature in this chapter.

For each choice of signature σ there is a dialect LP(σ) of LP.

Definition 3.1.1 For each signature σ:

(a) The *lexicon* of LP(σ) is the set of symbols consisting of the truth function symbols (3.1), the parentheses (3.2) and the symbols in σ.

(b) An *expression* of LP(σ) is a string of one or more symbols from the lexicon of LP(σ). The *length* of the expression is the number of occurrences of symbols in it. (Often the same symbol will occur more than once.)

It will be useful to be able to say things like 'If ϕ is an expression then so is $(\neg\phi)$', where the Greek letter ϕ is not an expression itself but a variable ranging over expressions.

Definition 3.1.2 When we are studying a language L, we distinguish between (1) symbols of L and (2) symbols like ϕ above, that are not in L but are used to range over expressions of L. The symbols in (2) are called *metavariables*. They will usually be Greek letters such as ϕ, ψ, χ.

We need to say which of the expressions of LP(σ) are 'grammatical sentences' of the language. These expressions are known as *formulas*. There is a short and sweet definition, as follows.

(3.5)
- $\bot$ is a formula of LP(σ), and so is every symbol in σ.
- If ϕ is a formula of LP(σ) then so is $(\neg\phi)$.
- If ϕ and ψ are formulas of LP(σ), then so are $(\phi \wedge \psi)$, $(\phi \vee \psi)$, $(\phi \rightarrow \psi)$ and $(\phi \leftrightarrow \psi)$.
- Nothing else is a formula of LP(σ).

Definitions, such as (3.5), that describe a set by saying that certain things are in it and it is closed under certain operations are known as *inductive definitions*. They require some mathematical explanation, for example, to make sense of the last bullet point. Even with that explanation taken as read, this definition of 'formula' is not necessarily the most helpful. It presents the formulas as a set of strings, and it hides the fact that formulas (like the sentences of any language) have a grammatical structure that underpins the uses we make of them. So in this and the next section we will follow an approach that has become standard in the study of natural languages, taking the grammatical structure as fundamental.

We begin with a particular formula of LP(σ) where σ contains p:

(3.6) $(p \rightarrow (\neg(\neg p)))$

This formula comes from the statement proved in Example 2.6.1 of Section 2.6, by putting p in place of ϕ throughout. It has the form $(\phi \rightarrow \psi)$ with p for ϕ and $(\neg(\neg p))$ for ψ. We can write this analysis as a diagram:

(3.7)

The formula on the left in (3.7) cannot be analysed any further. But on the right, $(\neg(\neg p))$ has the form $(\neg\phi)$ where ϕ is $(\neg p)$. So we can extend the diagram downwards:

(3.8)

One more step analyses $(\neg p)$ as the result of negating p:

(3.9)

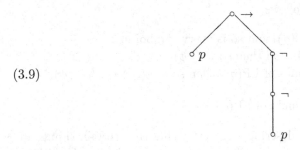

So we have a diagram (3.9) of small circles (we call them *nodes*) joined by lines, with labels on the right of the nodes. The labels on the two bottom nodes are one-symbol formulas (in this case both p). The other nodes carry the labels $\rightarrow$, $\neg$ and $\neg$. The diagram with its labels analyses how formula (3.6) was put together. This kind of grammatical analysis is traditionally known as *parsing*. The diagram branches downward, rather like a family tree, and so it will be called the *parsing tree* of the formula.

Given the tree diagram, we can reconstruct the formula that it came from. We illustrate this with a different tree that uses the signature $\{p_0, p_1\}$:

(3.10)

We reconstruct the formula by starting at the bottom tips of the tree and working upwards. As we go, we record our progress by writing labels on the *left* side of the tree nodes. The first step is to copy each propositional symbol on the left side of its node:

(3.11)

In the middle branch, the $\neg$ on the node just above p_0 shows that we form the formula $(\neg p_0)$. We write this formula on the left of the node:

(3.12)

Now the $\to$ on the left joins p_1 and $(\neg p_0)$ together to give $(p_1 \to (\neg p_0))$, and we attach this on the left of the node that carries the $\to$:

(3.13)

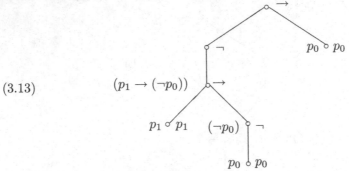

With two more moves to incorporate the remaining $\neg$ and $\to$, we finally reach the tree

(3.14)

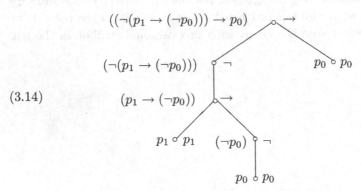

with $((\neg(p_1 \to (\neg p_0))) \to p_0)$ on the left of the top node. This is the formula constructed by the tree. If we started with this formula and decomposed it to form a tree, the result would be (3.10). So the tree and the formula go hand in hand; we say the formula is *associated with* the tree, and the tree is the *parsing tree* of the formula. The labels on the left sides of nodes in (3.14) form the *syntax labelling* of the parsing tree; the next section will define this more rigorously.

Pause for a moment to think about the way we constructed the labels on the left of the nodes. We started at the bottom of the tree and worked upwards. At each node we knew how to write the label on its left, depending only on the symbol written on its right and the left-hand labels on the nodes (if any) one level down. A set of instructions for writing a left labelling in this way is called a *compositional definition*. If δ is a compositional definition and we apply it to a parsing tree π, the thing we are trying to find is the left label on the top node; we call this label the *root label* and we write it $\delta(\pi)$.

In the example above, the labels on the left of nodes were expressions. But there are also compositional definitions that put other kinds of label on the left of nodes. For example, this is a compositional definition:

> Put 0 on the left of each node at the bottom of the tree. Next, if you
> have left labelled the nodes immediately below a given node ν, say
> with numbers $m_1, \ldots, m_n$, then write

$$(3.15) \qquad \max\{m_1, \ldots, m_n\} + 1$$

> on the left of ν.

This definition does not involve labels on the right at all. The label that it puts
on the left of a node is called the *height* of the node; the root label is called the
height of the tree.

Remark 3.1.3 Parsing trees were first designed by Frege in 1879 for his
Begriffsschrift. Here is Frege's own version of (3.10):

$$(3.16)$$

Our top is his top left. His signature used a and b instead of our p_0 and p_1.
He marked $\neg$ with a short line jutting out downwards, and $\rightarrow$ with an unla-
belled branching. Frege himself used his parsing trees as formulas. The problem
with this approach is that you cannot pronounce a parsing tree aloud, and even
writing it can use up paper and patience. Imagine writing out a natural deduction derivation with parsing trees as formulas! In this book we follow Frege in
using parsing trees to show the structure of formulas, but we do not try to use
them in place of formulas.

Exercises

3.1.1. For each of the following formulas, draw a parsing tree that has the formula as its associated formula. (Trial and error should suffice, but later
we will give an algorithm for this.)

(a) $(p \wedge q)$.

(b) p.

(c) $(p \rightarrow (q \rightarrow (r \rightarrow s)))$.

(d) $((\neg(p_2 \rightarrow (p_1 \leftrightarrow p_0))) \vee (p_2 \rightarrow \bot))$,

(e) $((p_6 \wedge (\neg p_5)) \rightarrow (((\neg p_4) \vee (\neg p_3)) \leftrightarrow (p_2 \wedge p_1)))$.

(f) $(((\neg(p_3 \wedge p_7)) \vee (\neg(p_1 \vee p_2))) \rightarrow (p_2 \vee (\neg p_5)))$.

(g) $(((\neg(p_1 \rightarrow p_2)) \wedge (p_0 \vee (\neg p_3))) \rightarrow (p_0 \wedge (\neg p_2)))$.

3.1.2. Find the associated formula of each of the following parsing trees.

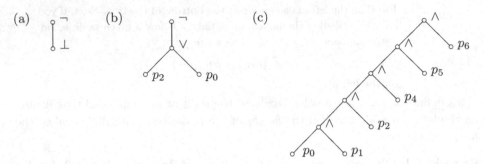

3.1.3. For each of the formulas in Exercise 3.1.1, find a smallest possible signature σ such that the formula is in the language $\mathrm{LP}(\sigma)$.

3.2 Parsing trees

In this section we make precise the ideas of Section 3.1. Note that the language LP, as we have constructed it so far, consists of strings of symbols. In this chapter we have not yet attached any meanings to these symbols, and we will not until Section 3.5.

We will define the formulas of $\mathrm{LP}(\sigma)$ in terms of their parsing trees. So first we need to define 'tree'—or more precisely 'planar tree', because for our trees it is important how they are written on the page.

Definition 3.2.1 A *(planar) tree* is an ordered pair (N, D) where

(a) N is a finite non-empty set whose elements are called *nodes*;

(b) D is a function that takes each node μ in N to a sequence (possibly empty) of distinct nodes:

$$(3.17) \qquad\qquad D(\mu) = (\nu_1, \ldots , \nu_n)$$

the nodes $\nu_1, \ldots , \nu_n$ are called the *daughters* of μ, and μ is called the *mother* of $\nu_1, \ldots , \nu_n$;

(c) every node except one has exactly one mother; the exception is a node called the *root*, in symbols $\sqrt{}$, which has no mother;

(d) there are no cycles, that is, sequences

$$(3.18) \qquad\qquad \nu_1, \nu_2, \ldots , \nu_k \;\; (k > 1)$$

where $\nu_k = \nu_1$ and each ν_i with $1 \leqslant i < k$ has mother ν_{i+1}.

To draw a tree (N, D), we first draw a node for its root $\sqrt{\ }$. If $D(\sqrt{\ }) = \langle \mu_1, \ldots, \mu_n \rangle$, then below the root we draw nodes $\mu_1, \ldots, \mu_n$ from left to right, joined to $\sqrt{\ }$ by lines. Then we put the daughters of μ_1 below μ_1, the daughters of μ_2 below μ_2, etc., and we carry on downwards until all the nodes are included. This will happen sooner or later, because N is finite by Definition 3.2.1(a); and if we start from any node and go to its mother, the mother of its mother and so on, then by (d) we must eventually reach a node with no mother, which by (c) must be $\sqrt{\ }$. The route from a node to the root is unique, since a node has at most one mother.

Definition 3.2.2

(a) In a tree, an *edge* is an ordered pair of nodes (μ, ν), where μ is the mother of ν. (So the lines in a tree diagram represent the edges.) Mixing metaphors, we describe a node as a *leaf* of a tree if it has no daughters. (Botanically speaking our trees are upside down, with their root at the top and their leaves at the bottom.)

(b) The number of daughters of a node is called its *arity*. (So the leaves are the nodes of arity 0.)

(c) We define a *height* for each node of a tree as follows. Every leaf has height 0. If μ is a node with daughters $\nu_1, \ldots, \nu_n$, then the height of μ is

$$(3.19) \qquad \max\{\text{height}(\nu_1), \ldots, \text{height}(\nu_n)\} + 1$$

The *height* of a tree is defined to be the height of its root (cf. (3.15)).

(d) A *path* from node ν to node μ is a set of nodes $\{\nu_0, \ldots, \nu_k\}$ where ν_0 is ν, ν_k is μ, and for each $i < k$, ν_i is the mother of ν_{i+1}. A path from the root to a leaf μ is called a *branch* (to μ).

Here are three examples of trees:

(3.20)

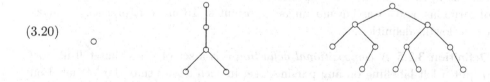

The left-hand tree in (3.20) has one leaf and no non-leaves, and its height is 0. The centre tree has two leaves, two nodes of arity 1 and one node of arity 2; the root has height 3, so the height of the tree is also 3. The right-hand tree has five leaves, two nodes of arity 1 and four nodes of arity 2; again the height of the tree is 3.

(3.21)

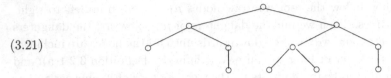

The tree (3.21) is strictly not the same tree as the third tree in (3.20), because the two trees are on different pages; also the angles are different in the two diagrams. But there is a unique correspondence between the nodes of one tree and the nodes of the other, and likewise between the edges of one tree and those of the others, which makes one tree a copy of the other (cf. Exercise 3.2.6). So it will do no harm if we 'identify' the two trees, that is, count them as the same tree.

Definition 3.2.3 A *labelling* of a tree is a function f defined on the set of nodes. We make it a *right labelling* by writing $f(\nu)$ to the right of each node ν, and a *left labelling* by writing $f(\nu)$ to the left of ν. A *labelled tree* is a tree together with a labelling; likewise we talk of *left-labelled trees* and *right-labelled trees*. (Sometimes we refer to right labels as *right-hand labels*, to avoid confusion between right/left and right/wrong.)

Definition 3.2.4 A *parsing tree* for LP(σ) is a right-labelled tree where

- every node has arity $\leqslant 2$;
- every leaf is labelled with either $\perp$ or a symbol from σ;
- every node of arity 1 is labelled with $\neg$;
- every node of arity 2 is labelled with one of $\wedge, \vee, \rightarrow; \leftrightarrow$.

Now you can easily check that all the parsing trees of Section 3.1 are parsing trees in the sense of Definition 3.2.4. So we can define the formulas of LP(σ) by saying how they are constructed from parsing trees. The method that we use, working up from leaves to the root as in Section 3.1, will have many applications. For example, we use it to set up alternative notations, and to define properties of formulas, and to assign meanings to formulas. To make it precise, we make the following definition.

Definition 3.2.5 A *compositional definition* δ is a set of rules that tell us how to put a left labelling on any parsing tree, in such a way that the left label on any node μ—write it $\delta(\mu)$—is determined by just two things:

- the right-hand label on μ and
- the sequence of values $(\delta(\nu_1), \dots, \delta(\nu_n))$ where μ has daughters $\nu_1, \dots, \nu_n$ from left to right.

(In parsing trees for LP, n can only be 0, 1 or 2.) The rules must always determine $\delta(\mu)$ uniquely from this data, so that they define a unique left labelling for any parsing tree π; the label on the root of π is called the *root label*, in symbols $\delta(\pi)$.

Example 3.2.6 The rules that we used in Section 3.1 for recovering a formula from its parsing tree form a compositional definition. For convenience we can write it

(3.22)

$$\chi \circ \chi \qquad \begin{array}{c}(\neg\phi) \circ \neg \\ \big| \\ \phi \circ\end{array} \qquad \begin{array}{c}(\phi\square\psi) \circ \square \\ \phi \circ \qquad \psi \circ\end{array}$$

where χ is $\bot$ or a propositional symbol in σ, and $\square$ is a truth function symbol $\wedge$, $\vee$, $\rightarrow$ or $\leftrightarrow$.

This is three instructions. The first says: at a leaf, copy the right-hand label on the left. The second says: at a node with right-hand label $\neg$, write $(\neg\phi)$ where ϕ is the left label on the daughter. The third tells you what to do at a node with two daughters. Together the three instructions cover all cases unambiguously.

Definition 3.2.7

(a) If π is a parsing tree for $\mathrm{LP}(\sigma)$, then the *formula associated to* π is $\delta(\pi)$ where δ is the compositional definition (3.22). We say that π is a *parsing tree for* $\delta(\pi)$. The *formulas* of $\mathrm{LP}(\sigma)$ are the formulas associated to parsing trees of $\mathrm{LP}(\sigma)$. A *formula* of LP is a formula of $\mathrm{LP}(\sigma)$ for some signature σ.

(b) The formula $\bot$ and propositional symbols are called *atomic* formulas. (These have parsing trees with just one node.) All other formulas are said to be *complex*.

(c) A formula has *complexity* k if it is associated to a parsing tree of height k. (So atomic formulas are those with complexity 0.)

For example, using the default signature, the following are atomic formulas:

$$\bot \qquad p_0 \qquad p_{2002} \qquad p_{999999999999}$$

(Remember that in LP, each of these counts as a single symbol, even though, for example, the last one is a string of length 13.) On the other hand the following three expressions

$$(\neg p_1) \qquad (p_0 \rightarrow (p_1 \rightarrow p_0)) \qquad ((p_1 \wedge (\neg p_0)) \leftrightarrow p_5)$$

are complex formulas. (It is easy to draw parsing trees for them.)

WARNING: Suppose the same formula was associated to two different parsing trees, one of height 17 and the other of height 18; what would the complexity of the formula be, according to Definition 3.2.7(c)? In fact this strange situation never occurs, but we should prove that it never does. That will be the task of Section 3.3.

Remark 3.2.8 Let π be a parsing tree and ν a node of π. Then removing all the nodes of π which are not ν and cannot be reached from ν by going downwards along edges, we get a smaller parsing tree. The left label on ν given by (3.22) is the label on the root of this smaller tree, so it is itself a formula. Thus all the left labels on nodes of a parsing tree given by (3.22) are formulas.

The next notion, though very useful, is rather hard to formalise. In practice there are indirect ways of formalising most of the information we get out of it. For the moment we will fall back on an informal description.

Imagine we are travelling up a parsing tree, using (3.22) to attach formulas on the left sides of nodes. Nothing gets thrown away: if we write a formula ϕ against a node, then ϕ reappears in the label of its mother node. In fact we can say what *part* of the label on the mother node is this ϕ, and we call this the *trace* of ϕ. Take, for example, a part of a parsing tree:

(3.23)

$$(\phi \vee \phi) \quad \vee$$

$$\phi \qquad \phi$$

We labelled both of the two lower nodes ϕ. Then the upper node gets the label $(\phi \vee \phi)$. In this label the left-hand ϕ is the trace of the ϕ on the bottom left node, and the right-hand ϕ is the trace of the ϕ on the bottom right node. We say in this case that there are two *occurrences* of ϕ in the formula $(\phi \vee \phi)$; each trace is a separate occurrence. If the parsing tree contains further nodes above the top node in (3.23), then we can trace each of the ϕ's in the labels on these nodes too.

Definition 3.2.9 Let ϕ be the associated formula of a parsing tree P. Then the *subformulas* of ϕ are the traces in ϕ of the left labels of all the nodes in P. (The formula ϕ itself counts as the trace of the root label.)

For example, the formula $((\neg(p_1 \rightarrow (\neg p_0))) \rightarrow p_0)$ has the parsing tree

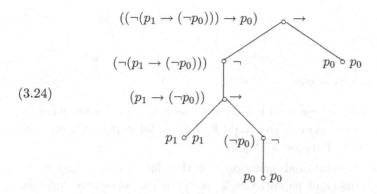

(3.24)

as we calculated in (3.14). The tree has seven nodes in its parsing tree, and hence the formula has seven subformulas, illustrated as follows:

$$((\neg(p_1 \rightarrow (\neg p_0))) \rightarrow p_0)$$

(3.25)

Note that the last two subformulas are the same formula, namely p_0, but occurring in different places.

Exercises

3.2.1.   List all the subformulas of the following formula. (You found its parsing tree in Exercise 3.1.1(d).)

$$((\neg(p_2 \rightarrow (p_1 \leftrightarrow p_0))) \vee (p_2 \rightarrow \bot)).$$

3.2.2.   Take σ to be the default signature $\{p_0, p_1, \dots\}$. Draw six parsing trees $\pi_1, \dots, \pi_6$ for $\mathrm{LP}(\sigma)$, so that each π_i has i nodes. Keep your parsing trees for use in later exercises of this section.

3.2.3. Consider the following compositional definition, which uses numbers as labels:

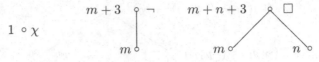

where χ is atomic and $\square \in \{\wedge, \vee, \rightarrow, \leftrightarrow\}$.

If π is any parsing tree for LP, and δ is the definition above, what is $\delta(\pi)$? Justify your answer. [You might find it helpful to try the definition in your trees from Exercise 3.2.2.]

3.2.4. Construct a compositional definition δ so that for each parsing tree π, $\delta(\pi)$ is the number of parentheses '(' or ')' in the associated formula of π.

3.2.5. This exercise turns formulas into numbers. Let σ be the default signature $\{p_0, p_1, p_2, \ldots\}$. We assign distinct odd positive integers $\sharp(s)$ to symbols s of $\mathrm{LP}(\sigma)$ as follows:

s	$\wedge$	$\vee$	$\rightarrow$	$\leftrightarrow$	$\neg$	$\perp$	p_0	p_1	$\ldots$
$\sharp(s)$	1	3	5	7	9	11	13	15	$\ldots$

The following compositional definition

(3.26) $2^{\sharp(\chi)} \circ \chi$

$$2^m \times 3^9 \overset{\neg}{\underset{m}{\bullet}} \qquad 2^m \times 3^n \times 5^{\sharp(\square)} \overset{\square}{\underset{m \quad n}{\diagup \diagdown}}$$

where χ is atomic and $\square \in \{\wedge, \vee, \rightarrow, \leftrightarrow\}$.

assigns a number to each node of any parsing tree. The number on the root is called the *Gödel number* of the associated formula of the tree. Explain how, if you know the Gödel number of a formula of $\mathrm{LP}(\sigma)$, you can reconstruct the formula. [Use unique prime decomposition.] Illustrate by reconstructing the formula with Gödel number

$$2^{2^{2^{15}} \times 3^9} \times 3^{2^{13}} \times 5^3$$

For goodness sake do not try to calculate the number!

3.2.6. Suppose (N_1, D_1) and (N_2, D_2) are planar trees. An *isomorphism* from (N_1, D_1) to (N_2, D_2) is a bijection $f : N_1 \rightarrow N_2$ such that for every node $\mu \in N_1$, if $D(\mu) = (\nu_1, \ldots, \nu_n)$ then $D(f\mu) = (f\nu_1, \ldots, f\nu_n)$. We say that two planar trees are *isomorphic* if there is an isomorphism from the first to the second. (Then isomorphism is an equivalence relation.) Prove: If f and g are two isomorphisms from (N_1, D_1) to (N_2, D_2) then $f = g$. [If (N_1, D_1) has height n, prove by induction on k that for each k, the functions f and g agree on all nodes of height $n - k$ in N_1.]

3.3 Propositional formulas

The main aim of this section is to prove that each formula of $\mathrm{LP}(\sigma)$ is associated to a unique parsing tree. When we analysed the formula $(p \rightarrow (\neg(\neg p)))$ in Section 3.1, we built a parsing tree for it. What has to be shown is that this method of analysis always works in a unique way, and it recovers the parsing tree that the formula is associated to.

By Definition 3.2.7 every formula ϕ of LP is the root label got by applying the compositional definition (3.22) to some parsing tree π. Looking at the clause of (3.22) used at the root, we see that ϕ is either atomic or a complex formula that can be written in at least one of the forms $(\neg\phi)$, $(\phi \wedge \psi)$, $(\phi \vee \psi)$, $(\phi \rightarrow \psi)$ or $(\phi \leftrightarrow \psi)$, where ϕ and ψ are formulas (using Remark 3.2.8). In the complex cases the shown occurrence of $\neg$, $\wedge$, $\vee$, $\rightarrow$ or $\leftrightarrow$ is said to be a *head* of the formula. To show that each formula has a unique parsing tree, we need to prove that each complex formula has a unique head. In fact this is all we need prove, because the rest of the tree is constructed by finding the heads of the formulas on the daughter nodes, and so on all the way down. Our proof will give an algorithm, that is, a mechanical method of calculation, for finding the head.

Purely for this section, we call the symbols $\neg, \wedge, \vee, \rightarrow, \leftrightarrow$ the *functors*. So a functor is a truth function symbol but not $\perp$.

Definition 3.3.1 As in Definition 3.1.1(b), an expression is a string $a_1 \cdots a_n$ of symbols and its length is n. A *segment* of this expression is a string

$$a_i \cdots a_j \quad (\text{with } 1 \leqslant i \leqslant j \leqslant n)$$

This segment is *initial* if $i = 1$; so the initial segments are

$$a_1 \qquad a_1 a_2 \qquad a_1 \cdots a_3 \qquad \cdots \qquad a_1 \cdots a_n$$

The *proper initial segments* of the expression are those of length $< n$. For each initial segment s we define the *depth* $d[s]$ to be the number of occurrences of '(' in s minus the number of occurrences of ')' in s. The *depth* of an occurrence a_j of a symbol in the string is defined as the depth of $a_1 \cdots a_j$.

Once again, remember that a propositional symbol is a single symbol. For example, we count p_{31416} as one symbol, not as a string of six symbols.

Example 3.3.2 In the string $(p_0 \to (p_1 \to p_0))$ the depths of the initial segments are as follows:

Initial segment	Depth
(	1
$(p_0$	1
$(p_0 \to$	1
$(p_0 \to ($	2
$(p_0 \to (p_1$	2
$(p_0 \to (p_1 \to$	2
$(p_0 \to (p_1 \to p_0$	2
$(p_0 \to (p_1 \to p_0)$	1
$(p_0 \to (p_1 \to p_0))$	0

Lemma 3.3.3 *Let χ be any formula of LP. Then*

(a) χ has depth 0, and every proper initial segment of χ has depth > 0;

(b) if χ is complex then exactly one occurrence of $\wedge$, $\vee$, $\to$, $\leftrightarrow$ or $\neg$ in χ has depth 1, and this occurrence is the unique head of χ.

Proof By Definition 3.2.7, χ is the associated formula of some parsing tree π. If μ is a node of π, write $\overline{\mu}$ for the formula assigned by (3.22) as left label of μ. We show that for every node μ of π, $\overline{\mu}$ has the properties (a) and (b) of the lemma. The proof is by induction on the height of μ.

Case 1: μ has height 0. Then μ is a leaf, so $\overline{\mu}$ is atomic, its depth is 0 and it has no proper initial segments.

Case 2: μ has height $k+1 > 0$, assuming the result holds for all nodes of height $\leqslant k$. Then $\overline{\mu}$ has one of the forms $(\phi \wedge \psi)$, $(\phi \vee \psi)$, $(\phi \to \psi)$, $(\phi \leftrightarrow \psi)$ or $(\neg\phi)$, depending on the right-hand label at μ. The cases are all similar; for illustration we take the first. In this case, μ has two daughters ν_1 and ν_2, and $\overline{\mu}$ is $(\overline{\nu_1} \wedge \overline{\nu_2})$. By induction assumption both $\overline{\nu_1}$ and $\overline{\nu_2}$ satisfy (a) and (b). The initial segments of $\overline{\mu}$ are as follows:

(α) (
   This has depth 1.

(β) $(s$ where s is a proper initial segment of $\overline{\nu_1}$.
   Since $\overline{\nu_1}$ satisfies (a), the depth $d[s]$ is at least 1, so the depth $d[(s]$ is at least 2.

(γ) $(\overline{\nu_1}$
   Since $\overline{\nu_1}$ satisfies (a), the depth is $1 + 0 = 1$.

(δ) $(\overline{\nu_1}\wedge$
   The depth is $1 + 0 + 0 = 1$; so the head symbol $\wedge$ has depth 1.

(ε) ($\overline{\nu_1} \wedge s$ where s is a proper initial segment of $\overline{\nu_2}$.
As in case (β), the fact that $\overline{\nu_1}$ and $\overline{\nu_2}$ satisfy (a) implies that the depth is at least 2.

(ζ) ($\overline{\nu_1} \wedge \overline{\nu_2}$
The depth is $1 + 0 + 0 + 0 = 1$.

(η) $\overline{\mu}$ itself
The depth is $1 + 0 + 0 + 0 - 1 = 0$ as required.

This proves that $\overline{\mu}$ satisfies (a). To prove that it satisfies (b), we note from (δ) that the head symbol has depth 1. If t is any other occurrence of a functor in $\overline{\mu}$, then t must be inside either $\overline{\nu_1}$ or $\overline{\nu_2}$, and it's not the last symbol since the last symbol of a complex formula is always ')'. Hence t is the end of an initial segment as in case (β) or (ε), and in both these cases the depth of t is at least 2. □

Theorem 3.3.4 (Unique Parsing Theorem) *Let χ be a formula of LP. Then χ has exactly one of the following forms:*

(a) χ is an atomic formula.

(b) χ has exactly one of the forms $(\phi \wedge \psi)$, $(\phi \vee \psi)$, $(\phi \rightarrow \psi)$, $(\phi \leftrightarrow \psi)$, where ϕ and ψ are formulas.

(c) χ has the form $(\neg \phi)$, where ϕ is a formula.

Moreover in case (b) the formulas ϕ and ψ are uniquely determined segments of ϕ. In case (c) the formula ϕ is uniquely determined.

Proof Every formula of LP has a parsing tree. As noted at the start of this section, from the possible forms in (3.22), it follows that every formula of LP has *at least* one of the forms listed in the theorem. It remains to prove the uniqueness claims.

Inspection shows whether χ is atomic. Suppose then that χ is complex. By the lemma, χ has a unique head. The first symbol of χ is '('. If the head is $\neg$, it is the second symbol of χ; if the head of χ is not $\neg$ then we are in case (b) and the second symbol of χ is the first symbol of ϕ, which cannot ever be $\neg$. So the second symbol determines whether we are in case (b) or (c).

In case (b) the occurrence of the head is uniquely determined as in the lemma. So ϕ is everything to the left of this occurrence, except for the first '(' of χ; and ψ is everything to the right of the occurrence, except for the last ')' of χ. Similarly in case (c), ϕ is the whole of χ except for the first two symbols and the last symbol. □

The theorem allows us to find the parsing tree of any formula of LP, starting at the top and working downwards.

Example 3.3.5 We parse $(p_1 \to ((\neg p_3) \vee \bot))$. It is clearly not atomic, so we check the depths of the initial segments in order to find the head. Thus

$$d[(] = 1, \quad d[(p_1] = 1, \quad d[(p_1 \to] = 1$$

Found it! The third symbol is a functor of depth 1, so it must be the head. Therefore, the formula was built by applying this head to p_1 on the left and $((\neg p_3) \vee \bot)$ on the right. The formula on the left is atomic. A check for the head of the formula on the right goes

$$d[(] = 1, \quad d[((] = 2, \quad d[((\neg] = 2, \quad d[((\neg p_3] = 2, \quad d[((\neg p_3)] = 1, \quad d[((\neg p_3)\vee] = 1$$

so again we have found the head. Then we need to find the head of $(\neg p_3)$. After this is done, we have analysed down to atomic formulas. So starting at the top, we can draw the complete parsing tree:

(3.27)

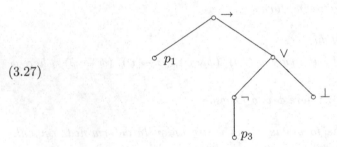

After your experience with Section 3.1, you could probably find this tree without needing the algorithm. But the algorithm makes the process automatic, and even experienced logicians can find this useful with complicated formulas.

Example 3.3.6 What happens if you try to use the algorithm above to find the parsing tree of an expression s that is not a formula? If you succeed in constructing a parsing tree, then s must be the associated formula of the tree, which is impossible. So you cannot succeed, but what happens instead? There is no question of going into an infinite loop; since the process breaks s down into smaller and smaller pieces, it has to halt after a finite time. What must happen is that you eventually hit an expression which is not an atomic formula and has no identifiable head. At this point the algorithm is telling you that s is not a formula.

For example, we try to parse $(p_0 \to p_1) \to p_2)$. The first $\to$ has depth 1, so we can get this far:

(3.28)

But at bottom right we need to calculate a tree for $p_1) \to p_2$. This expression is not an atomic formula and it does not contain any functor of depth 1. So the procedure aborts at this point and tells us that $(p_0 \to p_1) \to p_2)$ is not a formula of LP.

The Unique Parsing Theorem often allows us to rewrite compositional definitions in a simpler form. For example, the definition

(3.29)

$$1 \circ \chi \qquad \overset{m+3 \;\circ\; \neg}{\underset{m \;\circ}{\big\uparrow}} \qquad \overset{m+n+3 \;\circ\; \Box}{\underset{m \;\circ \qquad n \;\circ}{\wedge}}$$

where χ is atomic and $\Box \in \{\wedge, \vee, \to, \leftrightarrow\}$.

shakes down to the equivalent definition of a function f giving the left labels:

(3.30)
$$\begin{aligned}
f(\phi) &= 1 \quad \text{when } \phi \text{ is atomic} \\
f((\neg\phi)) &= f(\phi) + 3; \\
f((\phi\Box\psi)) &= f(\phi) + f(\psi) + 3 \qquad \text{when } \Box \in \{\wedge, \vee, \to, \leftrightarrow\}.
\end{aligned}$$

This second definition works because for each formula, exactly one case of the definition of f applies, and the formulas ϕ and ψ are uniquely determined. Because the definition of f needs only one line for each case, not a tree diagram, we say that this definition is got by *flattening* the compositional definition. Definitions, such as (3.30), that define some property of formulas by defining it outright for atomic formulas, and then for complex formulas in terms of their smaller subformulas, are said to be *recursive*, or *by recursion on complexity*. The name is because the same clause may recur over and over again. For example, in the calculation of $f(\phi)$, the clause for $\neg$ in (3.30) will be used once for each occurrence of $\neg$ in ϕ.

Exercises

3.3.1. Calculate the depths of all the initial segments of the string

$$(\neg(p_{22} \leftrightarrow (\neg\bot))).$$

3.3.2. Prove the case when $\bar{\mu}$ has the form $(\neg\phi)$ in Case Two of the proof of Lemma 3.3.3.

3.3.3. Calculate the heads of the following formulas of LP:
   (a) $((((\neg(\neg p_0)) \leftrightarrow (\neg(p_1 \to \bot))) \to p_1) \to p_1)$.
   (b) $(\neg((\neg p_0) \leftrightarrow ((((\neg p_1) \to \bot) \to p_1) \to p_1)))$.

3.3.4. True or false?: Given a formula ϕ of LP, if you remove the parentheses, then for every occurrence of $\wedge, \vee, \to$ or $\leftrightarrow$ in ϕ, there is a way of putting parentheses back into the formula so as to create a formula of LP in which the given occurrence is the head.

3.3.5. For each of the following strings, either write down its parsing tree (thereby showing that it is a formula of LP), or show by the method of Example 3.3.6 that it is not a formula of LP.

(a) $((p_3 \land (p_3 \land p_2)))$.

(b) $((((\neg p_1) \leftrightarrow \bot) \lor p_1) \land p_2)$.

(c) $(((\neg(p_0 \lor p_1)) \land (p_2 \to p_3))) \to (p_3 \land p_4))$.

(d) $(p_1 \land \neg\neg(p_2 \lor p_0))$.

(e) $((\neg p_1) \to (\neg p_2) \land p_1))$.

(f) $((p_1 \land (p_2 \lor p_3)) \leftrightarrow (\neg(\neg p_0)))$.

(g) $(((p_1 \land p_2)) \land (p_3 \land p_4))$.

(h) $((p_1 \to (\neg(p_2))) \leftrightarrow p_3)$.

(i) $(p_1 \to (p_2 \to (p_3 \land p_4) \to p_5)))$.

3.3.6. Item (a) below gives us a second method for showing that a given expression is not a formula of $LP(\sigma)$ for a given signature σ. Its advantage is that you normally do not need parsing trees for it. Its disadvantage is that you have to have an idea, unlike the method of Example 3.3.6 which is an algorithm that could be done by a computer.

(a) Prove the following: let S be a set of expressions such that

(1) every atomic formula of $LP(\sigma)$ is in S;

(2) if s and t are any expressions in S, then the expressions

$$(\neg s) \quad (s \land t) \quad (s \lor t) \quad (s \to t) \quad (s \leftrightarrow t)$$

are all in S.

Then every formula of $LP(\sigma)$ is in S.
[Let π be a parsing tree. Show (by induction on height of ν) that if (1) and (2) are true then for every node ν of π, the formula $\overline{\nu}$ at ν is in S.]

(b) Use (a) to show that every formula of $LP(\sigma)$ has equal numbers of left parentheses '(' and right parentheses ')'. [Put $S = \{s \mid s$ has equal numbers of left and right parentheses$\}$, and remember to prove that (1) and (2) are true for this S.] Deduce that

$$(((p \to q) \lor (\neg p))$$

is not a formula of $LP(\sigma)$.

3.3.7. In each case below, use Exercise 3.3.6(a) to show that the given expression is not a formula of $\mathrm{LP}(\sigma)$ where $\sigma = \{p_0, p_1, \dots\}$, by finding a set that satisfies (1) and (2) above but does not contain the given expression. [The expressions all do have equal numbers of left and right parentheses, so you cannot just use S from (b) of the previous exercise. Avoid mentioning 'formulas' in the definition of your S.]

(a) $p_{\sqrt{2}}$.

(b) $)p_0($.

(c) $(p_1 \wedge p_2 \to p_3)$.

(d) $(\neg\neg p_1)$. [WARNING. The obvious choice $S = \{s \mid s$ does not have two $\neg$ next to each other$\}$ does not work, because it fails (2); $\neg$ is in S but $(\neg\neg)$ is not.]

(e) $(p_1 \to ((p_2 \to p_3)) \to p_2)$.

(f) $(\neg p_1)p_2$.

(g) $(\neg p_1 \to (p_1 \vee p_2))$.

3.3.8. The Unique Parsing Theorem makes essential use of the parentheses '(' and ')' in a formula. But there are other logical notations that do not need parentheses. For example, *Polish* (also known as *head-initial*) notation has the following compositional definition:

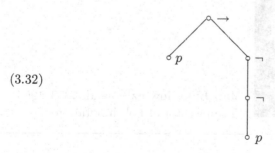

(3.31)

where χ is atomic, $\wedge'$ is K, $\vee'$ is A, $\to'$ is C and $\leftrightarrow'$ is E.

(a) Verify that in Polish notation the associated formula of the parsing tree

(3.32)

is $CpNNp$.

(b) Construct the parsing trees of the following Polish-notation formulas:

(i) $ENpq$.

(ii) $CCCpppp$.

(iii) $CNpAqKpNq$.

(c) Translate the following formulas from LP notation to Polish notation:

(i) $(p \vee (q \wedge (\neg p)))$.

(ii) $(((p \rightarrow q) \rightarrow p) \rightarrow p)$.

(d) Translate the following formulas from Polish notation to LP notation:

(i) $EApqNKNpNq$.

(ii) $CCqrCCpqCpr$.

3.3.9. Formulate a Unique Parsing Theorem for LP with Polish notation (cf. Exercise 3.3.8), and prove your theorem. [The problem is to show, for example, that if $K\phi\psi$ and $K\phi'\psi'$ are the same formula then $\phi = \phi'$ and $\psi = \psi'$. Try using a different depth function d defined by: $d[s] =$ the number of symbols from K, A, C, E in s, minus the number of propositional symbols in s, plus 1.] If the examples in Exercise 3.3.8 are not enough, you can test your algorithm and your eyesight on the following sentence in a book by Łukasiewicz:

$$CCCpqCCqrCprCCCCqrCprsCCpqs.$$

Jan Łukasiewicz Poland, 1878–1956.
The inventor of Polish notation.

3.3.10. (a) We write $NS(\phi)$ for the number of occurrences of subformulas in ϕ. Write out a compositional definition for $NS(\phi)$, and then flatten it to a recursive definition of $NS(\phi)$. ($NS(\phi)$ is equal to the number of nodes in the parsing tree for ϕ, so your compositional definition can be a slight adjustment of Exercise 3.2.3.)

(b) Write a recursive definition for Sub(ϕ), the set of formulas that occur as subformulas of ϕ. (There is a compositional definition for Sub(ϕ), but in fact a recursive definition is easier to write down directly.)

3.4 Propositional natural deduction

Intuitively speaking, derivations in this chapter are the same thing as derivations in Chapter 2, except that now we use formulas of LP instead of English statements. But we need to be more precise than this, for two reasons. First, we want to be sure that we can check unambiguously whether a given diagram is a derivation or not. Second, we need a description of derivations that will support our later mathematical analysis (e.g. the Soundness proof in Section 3.9, or the general results about provability in Chapter 8).

Our starting point will be the fact that the derivations of Chapter 2 have a tree-like feel to them, except that they have their root at the bottom and they branch upwards instead of downwards. (From a botanical point of view, of course, this is correct.) So we can borrow the definitions of Section 3.2, but with the trees the other way up. We will think of derivations as a kind of left-and-right-labelled tree, and we will define exactly which trees we have in mind. In practice we will continue to write derivations in the style of Chapter 2, but we can think of these derivations as a kind of shorthand for the corresponding trees.

To illustrate our approach, here is the derivation of Example 2.4.5:

$$(3.33) \qquad \cfrac{\cfrac{\phi^{①} \qquad (\phi \rightarrow \psi)}{\psi}\ (\rightarrow E) \qquad (\psi \rightarrow \chi)}{①\ \cfrac{\chi}{(\phi \rightarrow \chi)}\ (\rightarrow I)}\ (\rightarrow E)$$

Here is our tree version of it.

(3.34)

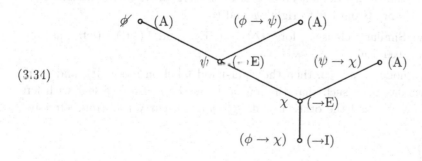

The formulas are the left labels. The right-hand label on a node tells us the rule that was used to bring the formula to the left of it into the derivation. Formulas not derived from other formulas are allowed by the Axiom Rule of Section 2.1, so we label them (A). Also we leave out the numbering of the discharged assumptions, which is not an essential part of the derivation.

With these preliminaries we can give a mathematical definition of 'derivation' that runs along the same lines as Definition 3.2.4 for parsing trees. The definition is long and repeats things we said earlier; so we have spelt out only the less obvious clauses. As you read it, you should check that the conditions in (d)–(g) correspond exactly to the natural deduction rules as we defined them in Chapter 2. (These rules are repeated in Appendix A.)

Definition 3.4.1 Let σ be a signature. Then a σ-*derivation* or, for short, a *derivation* is a left-and-right-labelled tree (drawn branching upwards) such that the following hold:

(a)   Every node has arity 0, 1, 2 or 3.

(b)   Every left label is either a formula of $LP(\sigma)$, or a formula of $LP(\sigma)$ with a dandah.

(c)   Every node of arity 0 carries the right-hand label (A).

(d)   If ν is a node of arity 1, then one of the following holds:

      (i) ν has right-hand label ($\rightarrow$I), and for some formulas ϕ and ψ, ν has the left label ($\phi \rightarrow \psi$) and its daughter has the left label ψ;

      (ii) ν has right-hand label ($\neg$I) or (RAA), the daughter of ν has left label $\perp$, and if the right-hand label on ν is ($\neg$I) then the left label on ν is of the form ($\neg\phi$).

  (iii),(iv),(v) Similar clauses for ($\wedge$E), ($\vee$I) and ($\leftrightarrow$E) (left as an exercise).

(e)   If ν is a node of arity 2, then one of the following holds:

      (i) ν has right-hand label ($\rightarrow$E), and there are formulas ϕ and ψ such that ν has the left label ψ, and the left labels on the daughters of ν are (from left to right) ϕ and ($\phi \rightarrow \psi$).

  (ii),(iii),(iv) Similar clauses for ($\wedge$I), ($\neg$E) and ($\leftrightarrow$I) (left as an exercise).

(f)   If ν is a node of arity 3, then the right-hand label on ν is ($\vee$E), and there are formulas ϕ, ψ such that the leftmost daughter of ν is a leaf with left label ($\phi \vee \psi$), and the other two daughters of ν carry the same left label as ν,

(g) If a node μ has left label χ with a dandah, then μ is a leaf, and the branch to μ (Definition 3.2.2(d)) contains a node ν where one of the following happens:

   (i) Case (d)(i) occurs with formulas ϕ and ψ, and ϕ is χ,

   (ii) Case (d)(ii) occurs; if the right-hand label on ν is ($\neg$I) then the left label on ν is ($\neg\chi$), while if it is (RAA) then χ is ($\neg\phi$) where ϕ is the left label on ν.

   (iii) ν has label ($\vee$E) with formulas ϕ and ψ as in Case (f), and either χ is ϕ and the path from the root to ν goes through the middle daughter of ν, or χ is ψ and the path goes through the right-hand daughter.

The *conclusion* of the derivation is the left label on its root, and its *undischarged assumptions* are all the formulas that appear without dandahs as left labels on leaves. The derivation is a *derivation of* its conclusion.

Theorem 3.4.2 *Let σ be a finite signature, or the signature $\{p_0, p_1, \ldots\}$. There is an algorithm that, given any diagram, will determine in a finite amount of time whether or not the diagram is a σ-derivation.*

Proof Definition 3.4.1 tells us exactly what to look for. The diagram must form a tree with both left and right labels, where every node has arity $\leqslant 3$. The left labels must all be formulas of LP(σ) (possibly with dandahs); Section 3.3 told us how to check this. The right labels must all be from the finite set ($\rightarrow$I), ($\rightarrow$E), etc. Each of the clauses (d)–(g) can be checked mechanically. $\qquad\square$

Leibniz would have been delighted. But in the seventeenth century, when Leibniz lived, it was assumed that any calculation must be with numbers. So when Leibniz asked for a way of calculating whether proofs are correct, he took for granted that this would involve converting the proofs into numbers; in fact he sketched some ideas for doing this. Exercise 3.2.5 was a modern version of the same ideas, due to Kurt Gödel. Since a derivation is also a tree, the same idea adapts and gives a Gödel number for each derivation. We will use this numbering in Chapter 8 to prove some important general facts about logic.

Example 3.4.3 Suppose D is a σ-derivation whose conclusion is $\bot$, and ϕ is a formula of LP(σ). Let D' be the labelled tree got from D by adding one new node below the root of D, putting left label ϕ and right label (RAA) on the new node, and writing a dandah on ($\neg\phi$) whenever it labels a leaf. We show that D' is a σ-derivation. The new node has arity 1, and its daughter is the root of D. Clearly (a)–(c) of Definition 3.4.1 hold for D' since they held for D. In (d)–(f) we need to only check for (d)(ii), the case for (RAA); D' satisfies this since the root of D carried $\bot$. There remains (g) with ($\neg\phi$) for χ: here D' satisfies (g)(ii), so the added dandahs are allowed. You will have noticed that we wrote D' as

(3.35)
$$
\frac{\begin{array}{c} \cancel{(\neg \phi)} \\ D \\ \bot \end{array}}{\phi} \;\; \text{(RAA)}
$$

in the notation of Chapter 2. We will continue to use that notation, but now we also have Definition 3.4.1 to call on when we need to prove theorems about derivations. (This example continues after Definition 3.4.4.)

We can now recast the definition of sequents, Definition 2.1.2, as follows.

Definition 3.4.4 Let σ be a signature. A σ-*sequent*, or for short just a *sequent*, is an expression

(3.36)
$$
\Gamma \vdash_\sigma \psi
$$

where ψ is a formula of $\mathrm{LP}(\sigma)$ (the *conclusion* of the sequent) and Γ is a set of formulas of $\mathrm{LP}(\sigma)$ (the *assumptions* of the sequent). The sequent (3.36) means

(3.37)   There is a σ-derivation whose conclusion is ψ and whose undischarged assumptions are all in the set Γ.

(This is a precise version of Definition (2.1.2).) When (3.37) is true, we say that the sequent is *correct*, and that the σ-derivation *proves* the sequent. The set Γ can be empty, in which case we write the sequent as

(3.38)
$$
\vdash_\sigma \psi
$$

This sequent is correct if and only if there is a σ-derivation of ψ with no undischarged assumptions.

When the context allows, we leave out σ and write $\vdash_\sigma$ as $\vdash$. This is innocent: Exercises 3.4.3 and 3.4.4 explain why the choice of signature σ is irrelevant so long as $\mathrm{LP}(\sigma)$ includes the assumptions and conclusion of the sequent.

Example 3.4.3 (continued)   Let Γ be a set of formulas of $\mathrm{LP}(\sigma)$ and ϕ a formula of $\mathrm{LP}(\sigma)$. We show that if the sequent $\Gamma \cup \{\neg\phi\} \vdash_\sigma \bot$ is correct, then so is the sequent $\Gamma \vdash_\sigma \phi$. Intuitively this should be true, but thanks to the definition (3.37) we can now prove it mathematically. By that definition, the correctness of $\Gamma \cup \{(\neg\phi)\} \vdash_\sigma \bot$ means that there is a σ-derivation D whose conclusion is $\bot$ and whose undischarged assumptions are all in $\Gamma \cup \{(\neg\phi)\}$. Now let D' be the derivation constructed from D earlier in this example. Then D' has conclusion ϕ and all its undischarged assumptions are in Γ, so it proves $\Gamma \vdash_\sigma \phi$ as required.

The Greek metavariables ϕ, ψ etc. are available to stand for any formulas of LP. For example, the derivation in Example 2.4.5 whose tree we drew is strictly not a derivation but a pattern for derivations. Taking σ as the default signature $\{p_0, p_1, \dots\}$, an example of a genuine σ-derivation that has this pattern is

$$
\cfrac{\cfrac{p_5^{①} \qquad (p_5 \to p_3)}{p_3}\ (\to E) \qquad (p_3 \to \bot)}{①\ \cfrac{\bot}{(p_5 \to \bot)}\ (\to I)}\ (\to E)
$$

This derivation is a proof of the sequent

$$\{(p_5 \to p_3), (p_3 \to \bot)\} \vdash_\sigma (p_5 \to \bot).$$

In practice, we will continue to give derivations using Greek letters, as a way of handling infinitely many derivations all at once.

Charles S. Peirce USA, 1839–1914.
A very inventive logician, one of the creators of modern semantics.

Problem: How do we know we have all the rules we ought to have for propositional natural deduction?

Example 3.4.5 The following example shows that this is not a fake problem. In the late nineteenth century, Charles Peirce showed the correctness of the sequent $\vdash (((\phi \to \psi) \to \phi) \to \phi)$. In fact, we can prove it by the following derivation:

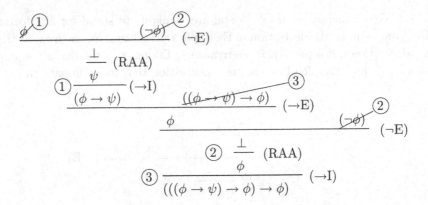

However, it is also possible to show that there is no derivation that proves this sequent and uses just the rules for → and the Axiom Rule (cf. Exercise 3.9.2). The derivation above needs the symbol ⊥ and the rule (RAA) as well. This raises the depressing thought that we might be able to prove still more correct sequents that use no truth function symbols except →, if we knew what extra symbols and rules to add.

Fortunately, the situation is not as bad as this example suggests. The class of acceptable rules of propositional logic is not open-ended; we can draw a boundary around it.

Recall that our language LP for propositional logic consists of meaningless statement-patterns, not actual statements. We can allow our propositional symbols to stand for any statements that we want, true or false as required. This gives us plenty of freedom for ruling out unwanted sequents: a sequent $\Gamma \vdash \psi$ is unacceptable if there is a way of reading the propositional symbols in it so that Γ becomes a set of truths and ψ becomes a falsehood.

Example 3.4.6 We show that the sequent $\{(p_0 \to p_1)\} \vdash p_1$ is unacceptable. To do this we *interpret* the symbols p_0 and p_1 by making them stand for certain sentences that are known to be true or false. The following example shows a notation for doing this:

(3.39)
$$\begin{array}{c|c} p_0 & 2 = 3 \\ p_1 & 2 = 3 \end{array}$$

Under this interpretation p_1 is false, but $(p_0 \to p_1)$ says 'If $2 = 3$ then $2 = 3$', which is true. So any rule which would deduce p_1 from $(p_0 \to p_1)$ would be unacceptable.

Definition 3.4.7 Let $(\Gamma \vdash \psi)$ be a σ-sequent, and let I be an interpretation that makes each propositional symbol appearing in formulas in the sequent into a meaningful sentence that is either true or false. Using this interpretation, each formula in the sequent is either true or false. (For the present, this is informal common sense; in the next section we will give a mathematical definition that allows us to calculate which formulas are true and which are false under a given interpretation.) We say that I is a *counterexample* to the sequent if I makes all the formulas of Γ into true sentences and ψ into a false sentence.

The moral of Example 3.4.6 is that if a sequent in propositional logic has a counterexample, then the sequent is unacceptable as a rule of logic. This suggests a programme of research: show that for every sequent of propositional logic, either there is a derivation (so that the sequent is correct in the sense of Definition 3.4.4) or there is a counterexample to it. If we can show this, then we will have shown that we have a complete set of rules for natural deduction; any new rules would either be redundant or lead to 'proofs' of unacceptable sequents. David Hilbert proposed this programme in lectures around 1920 (though in terms of a different proof calculus, since natural deduction hadn't yet been invented). The programme worked out well, and in sections 3.9 and 3.10 we will see the results.

David Hilbert Germany, 1862–1943.
Hilbert's Göttingen lectures of 1917–1922 were the first systematic presentation of first-order logic.

Exercises

3.4.1. Add the missing clauses (d)(iii)–(v) and (e)(ii)–(iv) to Definition 3.4.1.

3.4.2. Let σ be the default signature $\{p_0, p_1, \dots\}$. Neither of the following two diagrams is a σ-derivation. In them, find all the faults that you can, and state which clause of Definition 3.4.1 is violated by each fault. (Consider the diagrams as shorthand for labelled trees, as (3.33) is shorthand for (3.34).)

(a)
$$
\cfrac{
  \cfrac{
    \cfrac{
      \cfrac{\dfrac{p_0 \qquad \cancel{(p_0 \to \bot)}}{\bot}\ (\to E)}{}\ (\neg E)
    }{\ }
  }{\ }
}{\ }
$$

$$
\dfrac{\dfrac{p_0 \qquad \cancel{(p_0 \to \bot)}}{\bot}\ (\to E)}{}\ (\neg E)
$$

$$
\dfrac{\dfrac{p_1}{(p_1 \to p_0)}\ (\to I) \qquad \cancel{(\neg(p_1 \to p_0))}}{\bot}\ (\to E)
$$

$$
\dfrac{\dfrac{\bot}{\bot}\ (RAA)}{}
$$

$$
\dfrac{(\neg(p_0 \to p_1)) \qquad \qquad \dfrac{\bot}{\bot}\ (RAA)}{((\neg(p_0 \to p_1)) \to \bot)}\ (\to I)
$$

(b)
$$
\dfrac{
  \dfrac{
    \dfrac{\cancel{(\neg((\neg p_2) \vee q))} \qquad \dfrac{\cancel{(\neg p_2)}}{((\neg p_2) \vee q)}\ (\vee I)}{\bot}\ (\neg E)
  }{p_2}\ (\neg I)
  \qquad
  \dfrac{
    \dfrac{\dfrac{p_2 \qquad \cancel{(p_2 \to q)}}{q}\ (\to E)}{((\neg p_2) \vee q)}\ (\vee I) \qquad \cancel{(\neg((\neg p_2) \vee q))}}{\dfrac{\bot}{(\neg p_2)}\ (RAA)}\ (\neg I)
}{\dfrac{\bot}{((\neg p_2) \vee q)}\ (\neg I)}
$$

3.4.3. Let ρ and σ be signatures with $\rho \subseteq \sigma$.

    (a)   Show that every parsing tree for $LP(\rho)$ is also a parsing tree for $LP(\sigma)$.

    (b)   Show that every formula of $LP(\rho)$ is also a formula of $LP(\sigma)$.

    (c)   Show that every ρ-derivation is also a σ-derivation. [Use Definition 3.4.1.]

    (d)   Show that if $(\Gamma \vdash_\rho \psi)$ is a correct sequent then so is $(\Gamma \vdash_\sigma \psi)$. [Use Definition 3.4.4.]

3.4.4. Let ρ and σ be signatures with $\rho \subseteq \sigma$.

    (a)   Suppose D is a σ-derivation, and D' is got from D by writing $\bot$ in place of each symbol in D that is in σ but not in ρ. Show that D' is a ρ-derivation. [Use Definition 3.4.1.]

    (b)   Suppose Γ is a set of formulas of $LP(\rho)$ and ψ is a formula of $LP(\rho)$, such that the sequent $(\Gamma \vdash_\sigma \psi)$ is correct. Show that the sequent $(\Gamma \vdash_\rho \psi)$ is correct. [If D is a σ-derivation proving $(\Gamma \vdash_\sigma \psi)$, apply part (a) to D and note that this does not change the conclusion or the undischarged assumptions.]

3.4.5. Let σ be a signature, Γ a set of formulas of $LP(\sigma)$ and ϕ a sentence of $LP(\sigma)$. Show the following, using Definitions 3.4.1 and 3.4.4:

(a) The sequent $\Gamma \cup \{(\neg\phi)\} \vdash_\sigma \bot$ is correct if and only if the sequent $\Gamma \vdash_\sigma \phi$ is correct. (Left to right was proved in Example 3.4.3.)

(b) The sequent $\Gamma \cup \{\phi\} \vdash_\sigma \bot$ is correct if and only if the sequent $\Gamma \vdash_\sigma (\neg\phi)$ is correct.

3.4.6. We have stated several rules about correct sequents, and verified them informally. With our new formal definition of sequents we can prove them mathematically. Do so using Definition 3.4.1. (The formulas mentioned are all assumed to be in $LP(\sigma)$.)

(a) The Axiom Rule: If ψ is a formula in Γ then $(\Gamma \vdash_\sigma \psi)$ is correct.

(b) Monotonicity: If $\Gamma \subseteq \Delta$ and $(\Gamma \vdash_\sigma \psi)$ is correct, then $(\Delta \vdash_\sigma \psi)$ is correct. (This was part of Exercise 2.1.3, but now we can do it with proper precision.)

(c) The Transitive Rule: If $(\Delta \vdash_\sigma \psi)$ is correct and for every formula χ in Δ, $(\Gamma \vdash_\sigma \chi)$ is correct, then $(\Gamma \vdash_\sigma \psi)$ is correct.

(d) The Cut Rule: If $(\Gamma \vdash_\sigma \phi)$ is correct and $(\Gamma \cup \{\phi\} \vdash_\sigma \psi)$ is correct, then $(\Gamma \vdash_\sigma \psi)$ is correct.

[The proofs of (c) and (d) involve taking derivations and fitting them together to create a new derivation.]

3.4.7. (a) In some systems of logic (mostly constructive systems where 'true' is taken to mean 'provable') there is a rule

if $(\Gamma \vdash (\phi \vee \psi))$ is a correct sequent then at least one of $(\Gamma \vdash \phi)$ and $(\Gamma \vdash \psi)$ is also correct.

By giving a counterexample to a particular instance, show that this is unacceptable as a rule for LP. [Start by giving counterexamples for both the sequents $(\vdash p_0)$ and $(\vdash (\neg p_0))$.]

(b) Aristotle (Greece, fourth century BC), who invented logic, once said 'It is not possible to deduce a true conclusion from contradictory premises' (*Prior Analytics* 64b7). He must have meant something subtler, but his statement looks like the following sequent rule:

if $(\{\phi\} \vdash \psi)$ and $(\{(\neg\phi)\} \vdash \psi)$ are correct sequents, then so is the sequent $(\vdash (\neg\psi))$.

By giving a counterexample to a particular instance, show that this is unacceptable as a rule for LP.

3.4.8. One of the reasons for moving to derivations in a formal language is that if we use sentences of English, then even the most plausibly valid sequents

have irritating counterexamples. For instance, in Example 2.4.5 we proved
the sequent

$$\{(\phi \to \psi), (\psi \to \chi)\} \vdash (\phi \to \chi)$$

But the fourteenth century English logician Walter Burley proposed the
counterexample

ϕ is the statement   'I imply you are a donkey'
ψ is the statement   'I imply you are an animal'
χ is the statement   'I imply the truth'.

What goes wrong in Burley's example? Your answer should consider
whether we could construct a mathematical counterexample of this kind,
and how we can ensure that nothing like this example can appear in our
formal language.

3.5 Truth tables

Definition 3.5.1 We take for granted henceforth that truth and falsehood are
two different things. It will never matter exactly what things they are, but in
some contexts it is convenient to identity truth with the number 1 and falsehood
with 0. We refer to truth and falsehood as the *truth values*, and we write them
as T and F, respectively. The *truth value of* a statement is T if the statement is
true and F if the statement is false.

In Section 3.4, we suggested that we can give a formula ϕ of LP a truth
value by giving suitable meanings to the propositional symbols in ϕ. In fact, it
turns out that the truth value of ϕ depends only on the truth values given to
the propositional symbols, and not on the exact choice of meanings.

Example 3.5.2 We want to calculate the truth value of a statement $(\phi \wedge \psi)$ from
the truth values of ϕ and ψ. The following table shows how to do it:

(3.40)

ϕ	ψ	$(\phi \wedge \psi)$
T	T	T
T	F	F
F	T	F
F	F	F

The table has four rows corresponding to the four possible assignments of truth
values to ϕ and ψ.

The first row in (3.40) says that if ϕ is true and ψ is true, then $(\phi \wedge \psi)$ is true. The second says that if ϕ is true and ψ is false, then $(\phi \wedge \psi)$ is false; and so on down.

We can do the same for all the symbols $\vee$, $\rightarrow$, $\leftrightarrow$, $\neg$, $\perp$, using the meanings that we gave them when we introduced them in Chapter 2. The following table shows the result:

(3.41)

ϕ	ψ	$(\phi \wedge \psi)$	$(\phi \vee \psi)$	$(\phi \rightarrow \psi)$	$(\phi \leftrightarrow \psi)$	$(\neg\phi)$	$\perp$
T	T	T	T	T	T	F	F
T	F	F	T	F	F		
F	T	F	T	T	F	T	
F	F	F	F	T	T		

The only part of this table that may raise serious doubts is the listing for $\rightarrow$. In fact not all linguists are convinced that the table is correct for 'If ... then' in ordinary English. But the following argument confirms that the table for $\rightarrow$ does correspond to normal mathematical usage.

In mathematics we accept as true that

(3.42)   if p is a prime > 2 then p is odd.

In particular,

(3.43)   if 3 is a prime > 2 then 3 is odd. (If T then T.)

This justifies the T in the first row. But also

(3.44)   if 9 is a prime > 2 then 9 is odd. (If F then T.)

This justifies the T in the third row. Also

(3.45)   if 4 is a prime > 2 then 4 is odd. (If F then F.)

This justifies the T in the fourth row. There remains the second row. But everybody agrees that if ϕ is true and ψ is false then 'If ϕ then ψ' must be false.

The table (3.41) tells us, for example, that $(\phi \wedge \psi)$ is true if we already know that ϕ and ψ are both true. But we have to start somewhere: to calculate whether a formula χ is true, we usually need to start by finding out which of the propositional symbols in χ are true and which are false; then we can apply (3.41), walking up the parsing tree for χ.

The propositional symbols in χ are just symbols; they do not mean anything, so they are not in themselves either true or false. But in applications of propositional logic, the propositional symbols are given meanings so that they will be true or false. As mathematicians we are not concerned with how meanings are assigned, but we are interested in what follows if the symbols have somehow been given truth values. So our starting point is an assignment of truth values to propositional symbols.

Definition 3.5.3 Let σ be a signature. By a σ-*structure* we mean a function A with domain σ, that assigns to each symbol p in σ a truth value $A(p)$.

We call this object a 'structure' in preparation for Chapters 5 and 7, where the corresponding 'structures' are much closer to what a mathematician usually thinks of as a structure. For practical application and for comparison with the later structures, we note that if σ is $\{q_1, \ldots, q_n\}$ (where the symbols are listed without repetition), then we can write the structure A as a chart

$$\frac{q_1 \quad \cdots \quad q_n}{A(q_1) \quad \cdots \quad A(q_n)}$$

Every σ-structure A gives a truth value $A^\star(\chi)$ to each formula χ of $\mathrm{LP}(\sigma)$ in accordance with table (3.41). It is useful to think of $A^\star(\chi)$ as the truth value that χ has 'in A'. (Compare the way that the sentence 'The rain stays mainly on the plain' is true 'in Spain'.) We can calculate the value $A^\star(\chi)$ by climbing up the parsing tree of ϕ as follows.

Example 3.5.4 Let χ be $(p_1 \wedge (\neg(p_0 \to p_2)))$. Given the $\{p_0, p_1, p_2\}$-structure:

$$A: \quad \frac{p_0 \quad p_1 \quad p_2}{\mathrm{F} \quad \mathrm{T} \quad \mathrm{T}}$$

we calculate the truth value $A^\star(\chi)$ of χ thus. Here is the parsing tree of χ, with the leaves marked to show their truth values in A:

(3.46)

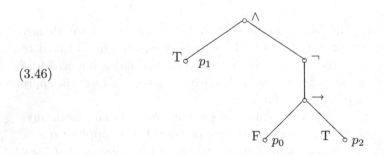

We work out the truth value at the node marked → by checking the third row of (3.41) for →: F → T makes T. So we label this node T:

(3.47)

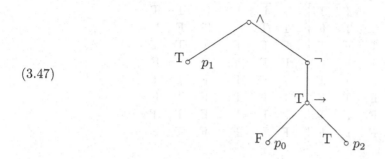

and so on upwards, until eventually we reach the root. This is marked F, so χ is false in A:

(3.48)

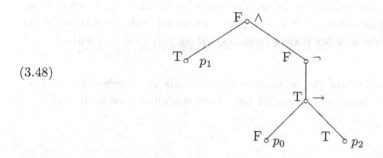

As always, the compositional definition attaches the left-hand labels, starting at the leaves of the parsing tree and working upwards.

The following table contracts (3.48) into two lines, together with an optional bottom line showing a possible order for visiting the nodes. The truth value at each node ν is written under the head of the subformula corresponding to ν. The head of χ itself is indicated by ⇑. We call the column with ⇑ the *head column* of the table; this is the last column visited in the calculation, and the value shown in it is the truth value of χ.

(3.49)

p_0	p_1	p_2	$(p_1$	$\wedge$	$(\neg$	$(p_0$	$\rightarrow$	$p_2)))$
F	T	T	T	F	F	F	T	T
			1	⇑	5	2	4	3

Table (3.49) shows the truth value of χ for one particular $\{p_0, p_1, p_2\}$-structure. There are eight possible $\{p_0, p_1, p_2\}$-structures. The next table lists them on the left, and on the right it calculates the corresponding truth value for χ. These values are shown in the head column.

(3.50)

p_0	p_1	p_2	$(p_1$	$\wedge$	$(\neg$	$(p_0$	$\rightarrow$	$p_2)))$
T	T	T	T	F	F	T	T	T
T	T	F	T	T	T	T	F	F
T	F	T	F	F	F	T	T	T
T	F	F	F	F	T	T	F	F
F	T	T	T	F	F	F	T	T
F	T	F	T	F	F	F	T	F
F	F	T	F	F	F	F	T	T
F	F	F	F	F	F	F	T	F

$\Uparrow$

Definition 3.5.5 A table like (3.50), which shows when a formula is true in terms of the possible truth values of the propositional symbols in it, is called the *truth table* of the formula. Note the arrangement: the first column, under p_0, changes slower than the second column under p_1, the second column changes slower than the third, and T comes above F. It is strongly recommended that you keep to this arrangement, otherwise other people (and very likely you yourself) will misread your tables.

Truth tables were invented by Charles Peirce in an unpublished manuscript of 1902, which may have been intended for a correspondence course in logic.

Models

Let σ be a signature and A a σ-structure. We have seen how A assigns a truth value $A^\star(\chi)$ to each formula χ of $\mathrm{LP}(\sigma)$. This function $A^\star$ is calculated by climbing up the parsing tree of χ, so it has a compositional definition; (B.3) in Appendix B shows how. But a flattened version by recursion on the complexity of χ is a little easier to write down (not least because it avoids boolean functions, which are explained in the Appendix). You should check that the parts of the following definition are in accordance with table (3.41).

Definition 3.5.6

(a) If p is a propositional symbol in σ then $A^\star(p) = A(p)$.

(b) $A^\star(\bot) = \mathrm{F}$.

(c) $A^\star((\neg\phi)) = \mathrm{T}$ if and only if $A^\star(\phi) = \mathrm{F}$.

(d) $A^\star((\phi \wedge \psi))$ is T if $A^\star(\phi) = A^\star(\psi) = \mathrm{T}$, and is F otherwise.

(e) $A^\star((\phi \vee \psi))$ is T if $A^\star(\phi) = \mathrm{T}$ or $A^\star(\psi) = \mathrm{T}$, and is F otherwise.

(f) $A^\star((\phi \rightarrow \psi))$ is F if $A^\star(\phi) = \mathrm{T}$ and $A^\star(\psi) = \mathrm{F}$, and is T otherwise.

(g) $A^\star((\phi \leftrightarrow \psi))$ is T if $A^\star(\phi) = A^\star(\psi)$, and is F otherwise.

Definition 3.5.7 Let σ be a signature, A a σ-structure and ϕ a formula of $\mathrm{LP}(\sigma)$. When $A^\star(\phi) = \mathrm{T}$, we say that A is a *model* of ϕ, and that ϕ is *true in* A. (In later chapters we will introduce the notation $\models_A \phi$ for 'A is a model of ϕ'.)

The process of checking whether a certain structure A is a model of a certain formula ϕ is known as *model checking*. This name is in use mostly among computer scientists, who have in mind commercial applications using much more intricate examples than our humble (3.49).

Several important notions are defined in terms of models.

Definition 3.5.8 Let σ be a signature and ϕ a formula of $\mathrm{LP}(\sigma)$.

(a)   We say that ϕ is *valid*, and that it is a *tautology*, in symbols $\models_\sigma \phi$, if every σ-structure is a model of ϕ. (So ($\models_\sigma \phi$) says that $A^\star(\phi) = \mathrm{T}$ for all σ-structures A.) When the context allows, we drop the subscript σ and write $\models \phi$.

(b)   We say that ϕ is *consistent*, and that it is *satisfiable*, if some σ-structure is a model of ϕ.

(c)   We say that ϕ is a *contradiction*, and that it is *inconsistent*, if no σ-structure is a model of ϕ.

If σ is the finite set $\{p_1, \ldots, p_n\}$, then we can check mechanically which of these properties ϕ has by calculating the truth table of ϕ, with all the $\{p_1, \ldots, p_n\}$-structures listed at the left-hand side. In fact ϕ is a tautology if and only if the head column of the table has T everywhere; ϕ is consistent if and only if the head column has T somewhere; and ϕ is a contradiction if and only if the head column has F everywhere.

Example 3.5.9 We confirm that Peirce's Formula, which we proved in Example 3.4.5, is a tautology:

(3.51)

p_1	p_2	$(((p_1$	$\rightarrow$	$p_2)$	$\rightarrow$	$p_1)$	$\rightarrow$	$p_1)$
T	T	T	T	T	T	T	T	T
T	F	T	F	F	T	T	T	T
F	T	F	T	T	F	F	T	F
F	F	F	T	F	F	F	T	F
							$\Uparrow$	

The notions in Definition 3.5.8 set us on track to use the ideas of Hilbert that we introduced at the end of Section 3.4. Simply put, the tautologies are the formulas that we ought to be able to prove. We will return to this in Sections 3.9 and 3.10, after exploring some important facts about tautologies.

In applications of the results of this section, the following fact is often used silently:

Lemma 3.5.10 (Principle of Irrelevance) *If σ is a signature, ϕ a formula of LP(σ) and A a σ-structure, then $A^\star(\phi)$ does not depend on the value of A at any propositional symbol that does not occur in ϕ.*

This is obvious: the rules for assigning $A^\star(\phi)$ never refer to $A(p)$ for any symbol p not occurring in ϕ. A formal proof would run as follows:

Let σ and τ be signatures, ϕ a formula of LP($\sigma \cap \tau$), and A and B a σ-structure and a τ-structure respectively. Suppose that $A(p) = B(p)$ for every propositional symbol $p \in \sigma \cap \tau$. Then we prove, by induction on the complexity of ϕ, that $A^\star(\phi) = B^\star(\phi)$.

Propositional logic is simple enough that the formal proof has no great advantage. But analogues of the Principle of Irrelevance apply also to more complicated languages; the more complex the language, the more important it becomes to check the Principle of Irrelevance formally.

Exercises

3.5.1. Prove by truth tables that the following are tautologies.

    (a) $(p \leftrightarrow p)$.

    (b) $(p \to (q \to p))$.

    (c) $((p_1 \to p_2) \leftrightarrow ((\neg p_2) \to (\neg p_1)))$.

    (d) $((p_1 \to (\neg p_1)) \leftrightarrow (\neg p_1))$.

    (e) $(p_1 \vee (\neg p_1))$.

    (f) $(\bot \to p_1)$.

    (g) $((p_1 \to (p_2 \to p_3)) \leftrightarrow ((p_1 \wedge p_2) \to p_3))$.

3.5.2. Write out truth tables for the formulas in the following list. For each of these formulas, say whether it is (a) a tautology, (b) a contradiction, (c) satisfiable. (It can be more than one of these.)

    (a) $((p_0 \to \bot) \leftrightarrow (\neg p_0))$.

    (b) $(p_1 \leftrightarrow (\neg p_1))$.

    (c) $((p_2 \wedge p_1) \to (\neg p_1))$.

    (d) $(((p_1 \leftrightarrow p_2) \wedge ((\neg p_1) \leftrightarrow p_3)) \wedge (\neg(p_2 \vee p_3)))$.

    (e) $((((p \to q) \to r) \to ((r \to p) \to q)) \to ((q \to r) \to p))$.

    (f) $((p_0 \to p_1) \to ((\neg(p_1 \wedge p_2)) \to (\neg(p_0 \wedge p_1))))$.

(g) $(((p \wedge q) \vee (p \wedge (\neg r))) \leftrightarrow (p \vee (r \to q)))$.

(h) $((p \wedge (\neg((\neg q) \vee r))) \wedge (r \vee (\neg p)))$.

3.5.3. You forgot whether the Logic class is at 11 or 12. Your friend certainly knows which; but sometimes he tells the truth and at other times he deliberately lies, and you know that he will do one of these but you do not know which. What should you ask him? [Let p be the statement that your friend is telling the truth, and let q be the statement that the lecture is at 11. You want to ask your friend whether a certain formula ϕ is true, where ϕ is chosen so that he will answer 'Yes' if and only if the lecture is at 11. The truth table of ϕ will be that of q if p is true, and that of $(\neg q)$ if p is false. Find an appropriate ϕ which contains both p and q.]

3.5.4. Let ρ and σ be signatures with $\rho \subseteq \sigma$, and let ϕ be a formula of LP(ρ). Explain how it follows from Lemma 3.5.10 (the Principle of Irrelevance) that $\models_\rho \phi$ if and only if $\models_\sigma \phi$.

3.5.5. Let σ be a signature containing k symbols. Calculate

(a) the number of σ-structures;

(b) the number $\alpha(k, n)$ of times that you need to write either T or F in writing out a truth table for a formula ϕ of LP(σ) which uses all of the propositional symbols in σ and has n nodes in its parsing tree;

(c) the largest value $\beta(\ell)$ of $\alpha(k, n)$ given that the formula ϕ has length ℓ.

The calculation of (c) shows that in the worst case, the size of a truth table of a formula of length ℓ rises exponentially with ℓ. Airline scheduling easily creates truth table problems with signatures of size greater than a million, and obviously for these one has to find some quicker approach if possible, depending on exactly what the question is.

3.6 Logical equivalence

Lemma 3.6.1 *Let σ be a signature and ϕ, ψ formulas of LP(σ). Then the following are equivalent:*

(i) *For every σ-structure A, A is a model of ϕ if and only if it is a model of ψ.*

(ii) *For every σ-structure A, $A^\star(\phi) = A^\star(\psi)$.*

(iii) $\models (\phi \leftrightarrow \psi)$.

Proof (i) and (ii) are equivalent by Definition 3.5.7.

(ii) and (iii) are equivalent by Definitions 3.5.6(g) and 3.5.8(a). $\square$

Definition 3.6.2 Let σ be a signature and ϕ, ψ formulas of LP(σ). We say that ϕ and ψ are *logically equivalent*, in symbols

$$\phi \quad \text{eq} \quad \psi$$

if any of the equivalent conditions (i)–(iii) of Lemma 3.6.1 hold.

Example 3.6.3 Clause (ii) in Lemma 3.6.1 says that ϕ and ψ have the same head column in their truth tables. We can use this fact to check logical equivalence. For example, the following truth table shows that

$$(p_1 \vee (p_2 \vee p_3)) \quad \text{eq} \quad ((p_1 \vee p_2) \vee p_3)$$

(3.52)

p_1	p_2	p_3	$(p_1$	$\vee$	$(p_2$	$\vee$	$p_3))$	$((p_1$	$\vee$	$p_2)$	$\vee$	$p_3)$
T	T	T	T	T	T	T	T	T	T	T	T	T
T	T	F	T	T	T	T	F	T	T	T	T	F
T	F	T	T	T	F	T	T	T	T	F	T	T
T	F	F	T	T	F	F	F	T	T	F	T	F
F	T	T	F	T	T	T	T	F	T	T	T	T
F	T	F	F	T	T	T	F	F	T	T	T	F
F	F	T	F	T	F	T	T	F	F	F	T	T
F	F	F	F	F	F	F	F	F	F	F	F	F
				⇑							⇑	

Theorem 3.6.4 *Let σ be a signature. Then eq is an equivalence relation on the set of all formulas of LP(σ). In other words it is*

- *Reflexive: For every formula ϕ, ϕ eq ϕ.*
- *Symmetric: If ϕ and ψ are formulas and ϕ eq ψ, then ψ eq ϕ.*
- *Transitive: If ϕ, ψ and χ are formulas and ϕ eq ψ and ψ eq χ, then ϕ eq χ.*

Proof   All three properties are immediate from Lemma 3.6.1(ii).   □

Example 3.6.5 Here follow some commonly used logical equivalences.

Associative Laws

$$(p_1 \vee (p_2 \vee p_3)) \quad \text{eq} \quad ((p_1 \vee p_2) \vee p_3)$$
$$(p_1 \wedge (p_2 \wedge p_3)) \quad \text{eq} \quad ((p_1 \wedge p_2) \wedge p_3)$$

Distributive Laws

$$(p_1 \vee (p_2 \wedge p_3)) \quad \text{eq} \quad ((p_1 \vee p_2) \wedge (p_1 \vee p_3))$$
$$(p_1 \wedge (p_2 \vee p_3)) \quad \text{eq} \quad ((p_1 \wedge p_2) \vee (p_1 \wedge p_3))$$

Commutative Laws

$$(p_1 \vee p_2) \quad \text{eq} \quad (p_2 \vee p_1)$$
$$(p_1 \wedge p_2) \quad \text{eq} \quad (p_2 \wedge p_1)$$

De Morgan Laws

$$(\neg(p_1 \vee p_2)) \quad \text{eq} \quad ((\neg p_1) \wedge (\neg p_2))$$
$$(\neg(p_1 \wedge p_2)) \quad \text{eq} \quad ((\neg p_1) \vee (\neg p_2))$$

Idempotence Laws

$$(p_1 \vee p_1) \quad \text{eq} \quad p_1$$
$$(p_1 \wedge p_1) \quad \text{eq} \quad p_1$$

Double Negation Law

$$(\neg(\neg p_1)) \quad \text{eq} \quad p_1$$

See also the equivalences in Exercise 3.6.2.

Exercises

3.6.1. Choose five of the equivalences in Example 3.6.5 (not including the first Associative Law) and prove them by truth tables.

3.6.2. Prove the following equivalences.

  (a) $(p \wedge q)$ is logically equivalent to $(\neg((\neg p) \vee (\neg q)))$, and to $(\neg(p \rightarrow (\neg q)))$.

  (b) $(p \vee q)$ is logically equivalent to $(\neg((\neg p) \wedge (\neg q)))$, and to $((p \rightarrow q) \rightarrow q)$.

  (c) $(p \rightarrow q)$ is logically equivalent to $(\neg(p \wedge (\neg q)))$, and to $((\neg p) \vee q)$.

  (d) $(p \leftrightarrow q)$ is logically equivalent to $((p \rightarrow q) \wedge (q \rightarrow p))$, and to $((p \wedge q) \vee ((\neg p) \wedge (\neg q)))$.

3.6.3. Show the following logical equivalences.

  (a) $(p_1 \leftrightarrow p_2)$ eq $(p_2 \leftrightarrow p_1)$.

  (b) $(p_1 \leftrightarrow (p_2 \leftrightarrow p_3))$ eq $((p_1 \leftrightarrow p_2) \leftrightarrow p_3)$.

  (c) $(\neg(p_1 \leftrightarrow p_2))$ eq $((\neg p_1) \leftrightarrow p_2)$.

  (d) $(p_1 \leftrightarrow (p_2 \leftrightarrow p_2))$ eq p_1.

3.6.4. Suppose ρ and σ are signatures with $\rho \subseteq \sigma$, and ϕ and ψ are formulas of LP(ρ). Show that ϕ and ψ are logically equivalent when regarded as formulas of LP(ρ) if and only if they are logically equivalent when regarded as formulas of LP(σ). [Use Lemma 3.6.1(iii) and Exercise 3.5.4.]

3.6.5. Show that the following are equivalent, for any formula ϕ of LP(σ):

  (a) ϕ is a tautology.

  (b) $(\neg\phi)$ is a contradiction.

(c) ϕ is logically equivalent to $(\neg\bot)$.

(d) ϕ is logically equivalent to some tautology.

3.7 Substitution

In this section we study what happens when we replace a part of a formula by another formula. We begin by making a double simplification. First, we limit ourselves to substitutions for propositional symbols. Second, we assume that a substitution changes either all or none of the occurrences of any given propositional symbol.

Definition 3.7.1 By a *substitution* S (for LP) we mean a function whose domain is a finite set $\{q_1, \dots, q_k\}$ of propositional symbols, and which assigns to each q_j $(1 \leqslant j \leqslant k)$ a formula ψ_i of LP. We normally write this function S as

(3.53) $\psi_1/q_1, \dots, \psi_k/q_k$

Changing the order in which the pairs ψ_i/q_i are listed does not affect the function. (To remember that it is ψ_1/q_1 and not q_1/ψ_1, think of ψ_1 as pushing down on q_1 to force it out of the formula.)

We apply the substitution (3.53) to a formula ϕ by simultaneously replacing every occurrence of each propositional symbol q_j in ϕ by ψ_j $(1 \leqslant j \leqslant k)$, and we write the resulting expression as $\phi[S]$, that is,

(3.54) $\phi[\psi_1/q_1, \dots, \psi_k/q_k]$

Example 3.7.2 Let ϕ be the formula

$$((p_1 \to (p_2 \wedge (\neg p_3))) \leftrightarrow p_3)$$

Let ψ_1 be $(\neg(\neg p_3))$, let ψ_2 be p_0 and let ψ_3 be $(p_1 \to p_2)$. Then the expression

$$\phi[\psi_1/p_1, \psi_2/p_2, \psi_3/p_3]$$

is

(3.55) $(((\neg(\neg p_3)) \to (p_0 \wedge (\neg(p_1 \to p_2)))) \leftrightarrow (p_1 \to p_2))$

The expression (3.55) is also a formula of LP, as we ought to expect. But from our explanation of (3.54) it is not immediately clear how one should prove that the expression $\phi[S]$ is always a formula of LP. So we need a more formal description of $\phi[S]$. To find one, we start from the fact that occurrences of a propositional symbol p correspond to leaves of the parsing tree that are labelled p.

A picture may help. Suppose we are constructing the expression $\phi[\psi/q]$. Let π be the parsing tree of ϕ, and $\nu_1, \ldots, \nu_n$ the leaves of π which are labelled q. Let τ be the parsing tree of ψ. Then we get a parsing tree of $\phi[\psi/q]$ by making n copies of τ, and fitting them below π so that the root of the i-th copy of τ replaces ν_i:

(3.56)

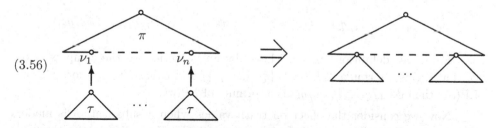

Now what happens when we climb up the new tree to find its associated formula? Starting at the bottom as always, we find the left labels on each of the copies of τ, and ψ will be the left label on the root of each of these copies. Then we label the rest of the tree; the process is exactly the same as when we label π *except that now the left labels on the nodes* $\nu_1, \ldots, \nu_n$ *are* ψ *and not* q. The situation with $\phi[\psi_1/q_1, \ldots, \psi_k/q_k]$ is very much the same but takes more symbols to describe.

In short, we build $\phi[\psi_1/q_1, \ldots, \psi_k/q_k]$ by applying a slightly altered version of (3.22), the definition of LP syntax, to the parsing tree of ϕ. The difference is the clause for leaf nodes, which now says

$$\chi' \circ \chi$$

(3.57)
$$\text{where } \chi' = \begin{cases} \bot & \text{if } \chi \text{ is } \bot, \\ \psi_i & \text{if } \chi \text{ is } q_i \ (1 \leqslant i \leqslant k), \\ p & \text{if } \chi \text{ is any other propositional symbol } p. \end{cases}$$

Flattening this compositional definition down gives the following recursive definition, which is our formal definition of $\phi[\psi_1/q_1, \ldots, \psi_k/q_k]$. Like our previous recursive definitions, it relies on the Unique Parsing Theorem.

Definition 3.7.3 Let $q_1, \ldots, q_k$ be propositional symbols, $\psi_1, \ldots, \psi_k$ formulas and ϕ a formula of LP. We define $\phi[\psi_1/q_1, \ldots, \psi_k/q_k]$ by recursion on the complexity of ϕ as follows.

If ϕ is atomic then

$$\phi[\psi_1/q_1, \ldots, \psi_k/q_k] = \begin{cases} \psi_i & \text{if } \phi \text{ is } q_i \ (1 \leqslant i \leqslant k), \\ \phi & \text{otherwise,} \end{cases}$$

If $\phi = (\neg\chi)$ where χ is a formula, then

$$\phi[\psi_1/q_1,\dots,\psi_k/q_k] = (\neg\chi[\psi_1/q_1,\dots,\psi_k/q_k])$$

If $\phi = (\chi_1\square\chi_2)$, where χ_1 and χ_2 are formulas and $\square \in \{\wedge,\vee,\rightarrow,\leftrightarrow\}$, then

$$\phi[\psi_1/q_1,\dots,\psi_k/q_k] = (\chi_1[\psi_1/q_1,\dots,\psi_k/q_k]\square\chi_2[\psi_1/q_1,\dots,\psi_k/q_k])$$

From this definition one can check, by an induction on the complexity of ϕ, that if ϕ is a formula of $\mathrm{LP}(\sigma \cup \{q_1,\dots,q_k\})$ and $\psi_1,\dots,\psi_k$ are formulas of $\mathrm{LP}(\sigma)$, then $\phi[\psi_1/q_1,\dots,\psi_k/q_k]$ is a formula of $\mathrm{LP}(\sigma)$.

Now we consider the effect on truth values when a substitution is made. The truth value $A^\star(\phi[\psi/q])$ of $\phi[\psi/q]$ in A is calculated by climbing up the tree in (3.56), just as the formula $\phi[\psi/q]$ itself was. The calculation is the same as for $A^\star(\phi)$, except that the truth value assigned to the nodes ν_i is $A^\star(\psi)$ instead of $A(q)$. So the effect is the same as if we calculated the truth value of ϕ in a structure $A[\psi/q]$, which is the same as A except that $A[\psi/q](q)$ is $A^\star(\psi)$. This motivates the following definition.

Definition 3.7.4 If A is a propositional structure and S is the substitution $\psi_1/q_1,\dots,\psi_k/q_k$, we define a structure $A[S]$ by

$$A[S](p) = \begin{cases} A^\star(\psi_j) & \text{if } p \text{ is } q_j \text{ where } 1 \le j \le k; \\ A(p) & \text{otherwise.} \end{cases}$$

Lemma 3.7.5 *Let A be a σ-structure and S the substitution $\psi_1/q_1,\dots,\psi_k/q_k$ with $\psi_1,\dots,\psi_k$ in $LP(\sigma)$. Then for all formulas ϕ of $LP(\sigma \cup \{q_1,\dots,q_k\})$,*

$$A^\star(\phi[S]) = A[S]^\star(\phi)$$

Proof We prove this by induction on the complexity of ϕ.
Case 1: ϕ has complexity 0. If ϕ is a propositional symbol that is not one of $q_1,\dots,q_k$, then

$$A^\star(\phi[S]) = A(\phi) = A[S]^\star(\phi)$$

If ϕ is q_i for some i then

$$A^\star(\phi[S]) = A^\star(\psi_i) = A[S]^\star(\phi)$$

If ϕ is $\bot$ then both $A^\star(\phi[S])$ and $A[S]^\star(\phi)$ are $\bot$.

Case 2: ϕ has complexity $k + 1$, assuming the lemma holds for all formulas of complexity $\leqslant k$. Suppose first that ϕ is $(\chi_1 \vee \chi_2)$ where χ_1 and χ_2 are formulas. Then

$$A^\star(\phi[S]) = \mathrm{T} \Leftrightarrow A^\star((\chi_1[S] \vee \chi_2[S])) = \mathrm{T}$$
$$\Leftrightarrow A^\star(\chi_1[S]) = \mathrm{T} \text{ or } A^\star(\chi_2[S]) = \mathrm{T}$$
$$\Leftrightarrow A[S]^\star(\chi_1) = \mathrm{T} \text{ or } A[S]^\star(\chi_2) = \mathrm{T}$$
$$\Leftrightarrow A[S]^\star(\phi) = \mathrm{T}$$

where the first step is by Definition 3.7.3, the second and final steps are by Definition 3.5.6(e), and the third step is by the induction hypothesis (since χ_1 and χ_2 have complexity at most k). It follows that $A^\star(\phi[S]) = A[S]^\star(\phi)$. The other possibilities under Case Two are that $\phi = (\chi_1 \,\square\, \chi_2)$ where $\square$ is $\wedge$, $\to$ or $\leftrightarrow$, and $\phi = (\neg\chi)$. These are similarly dealt with and left to the reader. $\square$

Lemma 3.7.5 allows us to increase our stock of tautologies and logical equivalences dramatically.

Theorem 3.7.6

(a) **(Substitution Theorem)** *Let S be a substitution and ϕ_1, ϕ_2 logically equivalent formulas of LP. Then $\phi_1[S]$ eq $\phi_2[S]$.*

(b) **(Substitution Theorem)** *Let S_1 and S_2 be the following substitutions:*

$$\psi_1/q_1, \ldots, \psi_k/q_k, \qquad \psi_1'/q_1, \ldots, \psi_k'/q_k$$

where for each j $(1 \leqslant j \leqslant k)$, ψ_j eq ψ_j'. Then for every formula ϕ, $\phi[S_1]$ eq $\phi[S_2]$.

Proof We assume that the formulas in question are in $\mathrm{LP}(\sigma)$, so that we can test logical equivalence by σ-structures.

(a) Let A be any σ-structure. Let $B = A[S]$ and note that by assumption, $B^\star(\phi_1) = B^\star(\phi_2)$. By two applications of Lemma 3.7.5,

$$A^\star(\phi_1[S]) = B^\star(\phi_1) = B^\star(\phi_2) = A^\star(\phi_2[S])$$

It follows that $\phi_1[S]$ eq $\phi_2[S]$.

(b) Again let A be any σ-structure. Since ψ_j eq ψ_j', $A^\star(\psi_j) = A^\star(\psi_j')$ for $1 \leqslant j \leqslant k$. Hence $A[S_1] = A[S_2]$, and by Lemma 3.7.5,

$$A^\star(\phi[S_1]) = A[S_1]^\star(\phi) = A[S_2]^\star(\phi) = A^\star(\phi[S_2])$$

It follows that $\phi[S_1]$ eq $\phi[S_2]$. $\square$

Example 3.7.7 of the Substitution Theorem. A formula is a tautology if and only if it is logically equivalent to $(\neg\bot)$ (cf. Exercise 3.6.5), and $(\neg\bot)[S]$ is just $(\neg\bot)$. Hence the Substitution Theorem implies that if ϕ is a tautology then so is $\phi[S]$. For example, the formula $(p \to (q \to p))$ is a tautology. By applying the substitution

$$(p_1 \wedge (\neg p_2))/p, \; (p_0 \leftrightarrow \bot)/q$$

we deduce that

$$((p_1 \wedge (\neg p_2)) \to ((p_0 \leftrightarrow \bot) \to (p_1 \wedge (\neg p_2))))$$

is a tautology. More generally, we could substitute any formula ϕ for p and any formula ψ for q, and so by the Substitution Theorem any formula of the form

$$(\phi \to (\psi \to \phi))$$

is a tautology.

Example 3.7.8 of the Replacement Theorem. We know that

(3.58) $(p_1 \wedge p_2) \; \text{eq} \; (\neg((\neg p_1) \vee (\neg p_2)))$

We would like to be able to put the right-hand formula in place of the left-hand one in another formula, for example, $((p_1 \wedge p_2) \to p_3)$. The trick for doing this is to choose another propositional symbol, say r, that does not occur in the formulas in front of us. (This may involve expanding the signature, but that causes no problems.) Then

$$((p_1 \wedge p_2) \to p_3) \quad \text{is} \quad (r \to p_3)[(p_1 \wedge p_2)/r]$$
$$((\neg((\neg p_1) \vee (\neg p_2))) \to p_3) \quad \text{is} \quad (r \to p_3)[(\neg((\neg p_1) \vee (\neg p_2)))/r]$$

Then the Replacement Theorem tells us at once that

$$((p_1 \wedge p_2) \to p_3) \; \text{eq} \; ((\neg((\neg p_1) \vee (\neg p_2))) \to p_3)$$

Example 3.7.9 of the Substitution Theorem. Starting with the same logical equivalence (3.58) as in the previous example, we can change the symbols p_1 and p_2, provided that we make the same changes in both formulas of (3.58). Let ϕ and ψ be any formulas, and let S be the substitution

$$\phi/p_1, \psi/p_2$$

Then $(p_1 \wedge p_2)[S]$ is $(\phi \wedge \psi)$, and $(\neg((\neg p_1) \vee (\neg p_2)))[S]$ is $(\neg((\neg\phi) \vee (\neg\psi)))$. So we infer from the Substitution Theorem that

$$(\phi \wedge \psi) \; \text{eq} \; (\neg((\neg\phi) \vee (\neg\psi))).$$

How to remember which is the Substitution Theorem and which the Replacement? In the **S**ubstitution Theorem it is the **S**ame **S**ubstitution on both **S**ides.

Exercises

3.7.1. Carry out the following substitutions.

    (a) $(p \to q)[p/q]$.

    (b) $(p \to q)[p/q][q/p]$.

    (c) $(p \to q)[p/q, q/p]$.

    (d) $(r \wedge (p \vee q))[((t \to (\neg p)) \vee q)/p, (\neg(\neg q))/q, (q \leftrightarrow p)/s]$.

3.7.2. (a) Using one of the De Morgan Laws (Example 3.6.5), show how the following equivalences follow from the Replacement and Substitution Theorems:

$$(\neg((p_1 \wedge p_2) \wedge p_3)) \quad \text{eq} \quad ((\neg(p_1 \wedge p_2)) \vee (\neg p_3))$$
$$\text{eq} \quad (((\neg p_1) \vee (\neg p_2)) \vee (\neg p_3)).$$

    (b) Show the following generalised De Morgan Law, by induction on n: If $\phi_1, \ldots, \phi_n$ are any formulas then

$$(\neg(\cdots (\phi_1 \wedge \phi_2) \wedge \cdots) \wedge \phi_n)) \quad \text{eq} \quad (\cdots ((\neg \phi_1) \vee (\neg \phi_2)) \vee \cdots) \vee (\neg \phi_n)).$$

    (And note that by the other De Morgan Law, the same goes with $\wedge$ and $\vee$ the other way round.)

    (c) The four formulas below are logically equivalent. Justify the equivalences, using the Replacement and Substitution Theorems, the De Morgan and Double Negation Laws (Example 3.6.5) and (a), (b) above.

      (i) $(\neg((p_1 \wedge (\neg p_2)) \vee (((\neg p_1) \wedge p_2) \wedge p_3)))$

      (ii) eq $((\neg(p_1 \wedge (\neg p_2))) \wedge (\neg(((\neg p_1) \wedge p_2) \wedge p_3)))$

      (iii) eq $(((\neg p_1) \vee (\neg(\neg p_2))) \wedge (((\neg(\neg p_1)) \vee (\neg p_2)) \vee (\neg p_3)))$

      (iv) eq $(((\neg p_1) \vee p_2) \wedge ((p_1 \vee (\neg p_2)) \vee (\neg p_3)))$

3.7.3. We know that $(p \to q)$ is logically equivalent to $((\neg p) \vee q)$, and hence (by the Substitution Theorem) for all formulas ϕ and ψ, $(\phi \to \psi)$ is logically equivalent to $((\neg \phi) \vee \psi)$.

    (a) Deduce that every formula of LP is logically equivalent to one in which $\to$ never occurs. [Using the Substitution and Replacement Theorems, show that if ϕ contains n occurrences of $\to$ with

$n > 0$, then ϕ is logically equivalent to a formula containing $n - 1$ occurrences of $\rightarrow$.]

(b)   Illustrate this by finding a formula that is logically equivalent to

$$((((p \rightarrow q) \rightarrow r) \leftrightarrow s) \rightarrow t)$$

in which $\rightarrow$ never appears.

3.7.4.   By Exercise 3.6.2, both $(p \wedge q)$ and $(p \vee q)$ are logically equivalent to formulas in which no truth function symbols except $\rightarrow$ and $\neg$ occur.

(a)   Show that each of $(p \leftrightarrow q)$ and $\perp$ is logically equivalent to a formula in which no truth function symbols except $\rightarrow$ and $\neg$ occur.

(b)   Find a formula of LP logically equivalent to

$$(q \vee (((\neg p) \wedge q) \leftrightarrow \perp))$$

in which no truth function symbols except $\rightarrow$ and $\neg$ occur.

3.7.5.   Let ϕ be a propositional formula using no truth function symbols except $\leftrightarrow$ and $\neg$. Show that ϕ is a tautology if and only if ϕ satisfies the following two conditions:

(a)   Every propositional symbol occurs an even number of times in ϕ.

(b)   The negation sign $\neg$ occurs an even number of times in ϕ.

[Exercise 3.6.3 should help.]

3.7.6.   Suppose S and T are substitutions. Show that there is a substitution ST such that for every formula ϕ,

$$\phi[ST] = \phi[S][T]$$

3.8   Disjunctive and conjunctive normal forms

So far we have been careful to write formulas of LP correctly according to the definition of the language. But some abbreviations are commonly used, and in this section they will help to make some formulas clearer.

Definition 3.8.1

(a)   A *conjunction* of formulas is a formula

(3.59) $(\cdots (\phi_1 \wedge \phi_2) \wedge \cdots) \vee \phi_n)$

where $\phi_1, \ldots, \phi_n$ are formulas; these n formulas are called the *conjuncts* of the conjunction. We allow n to be 1, so that a single formula is a conjunction of itself. We abbreviate (3.59) to

$$(\phi_1 \wedge \cdots \wedge \phi_n)$$

leaving out all but the outside parentheses.

(b) A *disjunction* of formulas is a formula

(3.60) $$(\cdots(\phi_1 \vee \phi_2) \vee \cdots) \vee \phi_n)$$

where $\phi_1, \ldots, \phi_n$ are formulas; these n formulas are called the *disjuncts* of the conjunction. We allow n to be 1, so that a single formula is a disjunction of itself. We abbreviate (3.60) to

$$(\phi_1 \vee \cdots \vee \phi_n)$$

leaving out all but the outside parentheses.

(c) The *negation* of a formula ϕ is the formula

(3.61) $$(\neg\phi)$$

We abbreviate (3.61) to

$$\neg\phi$$

A formula that is either an atomic formula or the negation of an atomic formula is called a *literal*.

It can be shown (though we will not show it) that even after these parentheses are left off, unique parsing still holds. In fact we can also safely leave off the outside parentheses of any complex formula, provided that it is not a subformula of another formula.

Remark 3.8.2 It is easily checked that (d) and (e) of Definition 3.5.6 generalise as follows:

(d) $A^\star(\phi_1 \wedge \cdots \wedge \phi_n) = \mathrm{T}$ if and only if $A^\star(\phi_1) = \cdots = A^\star(\phi_n) = \mathrm{T}$.

(e) $A^\star(\phi_1 \vee \cdots \vee \phi_n) = \mathrm{T}$ if and only if $A^\star(\phi_i) = \mathrm{T}$ for at least one i.

Definition 3.8.3 Let σ be a signature and ϕ a formula of LP(σ). Then ϕ determines a function $|\phi|$ from the set of all σ-structures to the set $\{\mathrm{T}, \mathrm{F}\}$ of truth values, by:

$$|\phi|(A) = A^\star(\phi) \quad \text{for each } \sigma\text{-structure } A$$

This function $|\phi|$ is really the same thing as the head column of the truth table of ϕ, if you read a T or F in the i-th row as giving the value of $|\phi|$ for the σ-structure described by the i-th row of the table. So the next theorem can be read as 'Every truth table is the truth table of some formula'.

Emil Post Poland and USA, 1897–1954.
'It is desirable ... to have before us the vision of
the totality of these [formulas] streaming
out ... through forms of ever-growing complexity'
(1921).

Theorem 3.8.4 (Post's Theorem) *Let σ be a finite non-empty signature and g
a function from the set of σ-structures to $\{T, F\}$. Then there is a formula ψ of
LP(σ) such that $g = |\psi|$.*

Proof Let σ be $\{q_1, \ldots, q_m\}$ with $m \geqslant 1$. We split into three cases.
Case 1: $g(A) = F$ for all σ-structures A. Then we take ψ to be $(q_1 \wedge \neg q_1)$, which
is always false.
Case 2: There is exactly one σ-structure A such that $g(A) = T$. Then take ψ
to be $q_1' \wedge \cdots \wedge q_m'$ where

$$q_i' = \begin{cases} q_i & \text{if} \quad A(q_i) = T, \\ \neg q_i & \text{otherwise.} \end{cases}$$

We write ψ_A for this formula ψ. Then for every σ-structure B,

$$
\begin{aligned}
|\psi_A|(B) = T \quad &\Leftrightarrow \quad B^\star(\psi_A) = T \\
&\Leftrightarrow \quad B^\star(q_i') = T \text{ for all } i \ (1 \leqslant i \leqslant m), \text{ by Remark 3.8.2} \\
&\Leftrightarrow \quad B(q_i) = A(q_i) \text{ for all } i \ (1 \leqslant i \leqslant m) \\
&\Leftrightarrow \quad B = A \\
&\Leftrightarrow \quad g(B) = T.
\end{aligned}
$$

So $|\psi_A| = g$.
Case 3: $g(A) = T$ exactly when A is one of $A_1, \ldots, A_k$ with $k > 1$. In this case
let ψ be $\psi_{A_1} \vee \cdots \vee \psi_{A_k}$. Then for every σ-structure B,

$$
\begin{aligned}
|\psi|(B) = T \quad &\Leftrightarrow \quad B^\star(\psi) = T \\
&\Leftrightarrow \quad B^\star(\psi_{A_j}) = T \text{ for some } j \ (1 \leqslant j \leqslant k), \text{ by Remark 3.8.2} \\
&\Leftrightarrow \quad B = A_j \text{ for some } j \ (1 \leqslant j \leqslant k) \\
&\Leftrightarrow \quad g(B) = T.
\end{aligned}
$$

So again $|\psi| = g$. $\square$

Example 3.8.5 We find a formula to complete the truth table

p_1	p_2	p_3	?
T	T	T	F
T	T	F	T
T	F	T	T
T	F	F	F
F	T	T	T
F	T	F	F
F	F	T	F
F	F	F	F

There are three rows with value T:

p_1	p_2	p_3	?	
T	T	T	F	
T	T	F	T	$\Leftarrow A_1$
T	F	T	T	$\Leftarrow A_2$
T	F	F	F	
F	T	T	T	$\Leftarrow A_3$
F	T	F	F	
F	F	T	F	
F	F	F	F	

The formula ψ_{A_1} is $p_1 \wedge p_2 \wedge \neg p_3$. The formula ψ_{A_2} is $p_1 \wedge \neg p_2 \wedge p_3$. The formula ψ_{A_3} is $\neg p_1 \wedge p_2 \wedge p_3$. So the required formula is

$$(p_1 \wedge p_2 \wedge \neg p_3) \vee (p_1 \wedge \neg p_2 \wedge p_3) \vee (\neg p_1 \wedge p_2 \wedge p_3)$$

Our proof of Post's Theorem always delivers a formula ψ in a certain form. The next few definitions will allow us to describe it.

Definition 3.8.6

- A *basic conjunction* is a conjunction of one or more literals, and a *basic disjunction* is a disjunction of one or more literals. A single literal counts as a basic conjunction and a basic disjunction.

- A formula is in *disjunctive normal form* (DNF) if it is a disjunction of one or more basic conjunctions.

- A formula is in *conjunctive normal form* (CNF) if it is a conjunction of one or more basic disjunctions.

Example 3.8.7

(1)

$$p_1 \wedge \neg p_1$$

is a basic conjunction, so it is in DNF. But also p_1 and $\neg p_1$ are basic disjunctions, so the formula is in CNF too.

(2)

$$(p_1 \wedge \neg p_2) \vee (\neg p_1 \wedge p_2 \wedge p_3)$$

is in DNF.

(3) Negating the formula in (2), applying the De Morgan Laws and removing double negations gives

$$\neg((p_1 \wedge \neg p_2) \vee (\neg p_1 \wedge p_2 \wedge p_3))$$
$$\text{eq} \quad \neg(p_1 \wedge \neg p_2) \wedge \neg(\neg p_1 \wedge p_2 \wedge p_3)$$
$$\text{eq} \quad (\neg p_1 \vee \neg\neg p_2) \wedge (\neg\neg p_1 \vee \neg p_2 \vee \neg p_3)$$
$$\text{eq} \quad (\neg p_1 \vee p_2) \wedge (p_1 \vee \neg p_2 \vee \neg p_3)$$

This last formula is in CNF. (See Exercise 3.7.2(c) for these equivalences.)

Theorem 3.8.8 *Let σ be a non-empty finite signature. Every formula ϕ of $LP(\sigma)$ is logically equivalent to a formula ϕ^{DNF} of $LP(\sigma)$ in DNF, and to a formula ϕ^{CNF} of $LP(\sigma)$ in CNF.*

Proof From the truth table of ϕ we read off the function $|\phi|$. The proof of Post's Theorem constructs a formula ψ of $LP(\sigma)$ such that $|\psi| = |\phi|$. By inspection, the formula ψ is in DNF. Now if A is any σ-structure, then

$$A^\star(\psi) = |\psi|(A) = |\phi|(A) = A^\star(\phi)$$

and hence ψ eq ϕ. So we can take ϕ^{DNF} to be ψ.

To find ϕ^{CNF}, first use the argument above to find $(\neg\phi)^{DNF}$, call it θ. Then θ eq $\neg\phi$, so $\neg\theta$ is logically equivalent to $\neg\neg\phi$ and hence to ϕ. Hence $\neg\theta$ is a formula of $LP(\sigma)$ which is logically equivalent to ϕ.

Now apply the method of Example 3.8.7(3) above to $\neg\theta$, pushing the negation sign $\neg$ inwards by the De Morgan Laws and then cancelling double negations, to get a logically equivalent formula in CNF. $\square$

Corollary 3.8.9 *Let σ be any signature (possibly empty). Every formula ϕ of $LP(\sigma)$ is logically equivalent to a formula of $LP(\sigma)$ in which no truth function symbols appear except $\wedge$, $\neg$ and $\bot$.*

Proof First suppose ϕ contains some propositional symbol, so that σ is not empty. The theorem tells us that ϕ is logically equivalent to a formula ϕ^{DNF} of

LP(σ) in which no truth function symbols appear except $\land$, $\lor$ and $\neg$. So we only need to get rid of the $\lor$. But we can do that by applying to ϕ^{DNF} the logical equivalence

(3.62) $\qquad\qquad\qquad (p \lor q) \quad \text{eq} \quad \neg(\neg p \land \neg q)$

with the help of the Replacement Theorem.

Next, if ϕ contains no propositional symbols, then ϕ must be built up from $\bot$ using truth function symbols. But every such formula has the value T or F, independent of any structure. So ϕ is logically equivalent to one of $\neg\bot$ and $\bot$. $\square$

Satisfiability of formulas in DNF and CNF

A formula in DNF is satisfiable if and only if at least one of its disjuncts is satisfiable. Consider any one of these disjuncts; it is a basic conjunction

$$\phi_1 \land \cdots \land \phi_m.$$

This conjunction is satisfiable if and only if there is a σ-structure A such that

$$A^\star(\phi_1) = \ldots = A^\star(\phi_m) = \text{T}$$

Since the ϕ_i are literals, we can find such an A unless there are two literals among $\phi_1, \ldots, \phi_n$ which are respectively p and $\neg p$ for the same propositional symbol p. We can easily check this condition by inspecting the formula. So checking the satisfiability of a formula in DNF and finding a model, if there is one, are trivial. (See Exercise 3.8.3(b) for a test of this.)

The situation with formulas in CNF is completely different. Many significant mathematical problems can be written as the problem of finding a model for a formula in CNF. The general problem of determining whether a formula in CNF is satisfiable is known as SAT. Many people think that the question of finding a fast algorithm for solving SAT, or proving that no fast algorithm solves this problem, is one of the major unsolved problems of twenty-first century mathematics. (It is the 'P = NP' problem.)

Example 3.8.10 A *proper m-colouring* of a map is a function assigning one of m colours to each country in the map, so that no two countries with a common border have the same colour as each other. A map is *m-colourable* if it has a proper m-colouring.

Suppose a map has countries $c_1, \ldots, c_n$. Write p_{ij} for 'Country c_i has the j-th colour'. Then finding a proper m-colouring of the map is equivalent to finding a model of a certain formula θ in CNF. Namely, take θ to be the conjunction of the following formulas:

(3.63)
$$p_{i1} \vee p_{i2} \vee \cdots \vee p_{im} \text{ (for all } i \text{ from 1 to } n);$$
$$\neg p_{ik} \vee \neg p_{jk} \text{ (for all } i, j, k \text{ where countries } c_i, c_j \text{ have a common border).}$$

More precisely, if A is a model of θ, then we can colour each country c_i with the first colour j such that $A^*(p_{ij}) = \text{T}$.

Exercises

3.8.1. For each of the following formulas ϕ, find a formula ϕ^{DNF} in DNF, and a formula ϕ^{CNF} in CNF, which are both equivalent to ϕ.

(a) $\neg(p_1 \to p_2) \vee \neg(p_2 \to p_1)$.

(b) $(p_2 \leftrightarrow (p_1 \wedge p_3))$.

(c) $\neg(p_1 \to p_2) \vee (p_0 \leftrightarrow p_2)$.

(d) $\neg(p_1 \wedge p_2) \to (p_1 \leftrightarrow p_0)$.

(e) $\neg(p \wedge q) \to (q \leftrightarrow r)$.

(f) $((p_0 \to p_1) \to p_2) \to (p_0 \wedge p_1)$.

(g) $((p \leftrightarrow q) \to r) \to q$.

(h) $(p_0 \to p_1) \to (\neg(p_1 \wedge p_2) \to \neg(p_0 \wedge p_2))$.

(i) $(p_1 \wedge p_2) \to \neg(p_3 \vee p_4)$.

(j) $p_1 \to (p_2 \to (p_3 \to p_4))$.

3.8.2. The formulas ϕ^{DNF} got by the method of Theorem 3.8.8 are not necessarily the most efficient. For example, consider the formula

$$(p_1 \wedge \neg p_2 \wedge p_3 \wedge \neg p_4) \vee (p_1 \wedge \neg p_2 \wedge \neg p_3 \wedge \neg p_4)$$

Here the two conjuncts are identical except that one has p_3 where as the other has $\neg p_3$. In this case we can leave out p_3; the whole formula is logically equivalent to $p_1 \wedge \neg p_2 \wedge \neg p_4$.

(a) Justify the statement above. [By the Distributive Law in Example 3.6.5, $(\phi \wedge \psi) \vee (\phi \wedge \neg \psi)$ is logically equivalent to $\phi \wedge (\psi \vee \neg \psi)$, and this in turn is logically equivalent to ψ.]

(b) Use this method to find a shorter formula in DNF that is logically equivalent to the following:

$$(p_1 \wedge \neg p_2 \wedge \neg p_3 \wedge p_4) \vee (p_1 \wedge \neg p_2 \wedge \neg p_3 \wedge \neg p_4) \vee (p_1 \wedge p_2 \wedge p_3 \wedge \neg p_4).$$

3.8.3. (a) For each of the following formulas in DNF, either find a model if there is one, or show that there is none. *Do not* write out truth tables for the formulas.

(i) $(p_1 \wedge p_2 \wedge \neg p_1) \vee (p_2 \wedge \neg p_3 \wedge p_3) \vee (\neg p_1 \wedge p_3)$.

(ii) $(p_2 \wedge \neg p_1 \wedge p_3 \wedge \neg p_5 \wedge \neg p_2 \wedge p_4) \vee (\neg p_2 \wedge p_1 \wedge \neg p_3 \wedge p_5 \wedge \neg p_8 \wedge \neg p_1)$

(b) In your own words, write instructions for your younger sister (who is not a mathematician) so that she can answer (i), (ii) and similar questions by herself.

3.8.4. Let σ be a signature containing k symbols. The relation eq splits the set of formulas of $LP(\sigma)$ into classes, where each class consists of a formula and all the other formulas logically equivalent to it. (These are the *equivalence classes* of the equivalence relation eq.) Calculate the number of equivalence classes of the relation eq. [By Post's Theorem, every possible head column in a truth table for σ describes an equivalence class.]

3.8.5. Let σ be a non-empty signature. Show that every formula of $LP(\sigma)$ is logically equivalent to a formula of $LP(\sigma)$ in which no truth function symbols are used except $\rightarrow$ and $\neg$.

3.8.6. Let σ be a non-empty signature. Let $LP^|$ be the result of expanding LP by adding a new truth function symbol $|$ (known as the *Sheffer stroke*) with the truth table

(3.64)

ϕ	ψ	$(\phi|\psi)$
T	T	F
T	F	T
F	T	T
F	F	T

Show that every formula of $LP^|(\sigma)$ is logically equivalent to a formula of $LP^|(\sigma)$ which contains no truth function symbols except $|$.

3.8.7. Show that $\neg p$ is not logically equivalent to any formula whose only truth function symbols are $\wedge$, $\vee$, $\rightarrow$ and $\leftrightarrow$. [Show that for any signature σ, if A is the structure taking every propositional symbol to T, and ϕ is built up using at most $\wedge$, $\vee$, $\rightarrow$ and $\leftrightarrow$, then $A^\star(\phi) = \text{T}$.]

3.9 Soundness for propositional logic

In Definition 3.4.4 of Section 3.4 we defined exactly what is meant by the sequent

(3.65) $$\Gamma \vdash_\sigma \psi$$

where σ is a signature, Γ is a set of formulas of LP(σ) and ψ is a formula of LP(σ). It will be convenient to write

$$\Gamma \nvdash_\sigma \psi$$

to express that (3.65) is *not* correct.

We also described a programme, proposed by Hilbert, for checking that the provable sequents (3.65) are exactly those that on grounds of truth and falsehood we ought to be able to prove. In this section and the next, we make the programme precise and carry it through for propositional logic.

Definition 3.9.1 Let σ be a signature, Γ a set of formulas of LP(σ) and ψ a formula of LP(σ).

(a) We say that a σ-structure A is a *model of* Γ if it is a model of every formula in Γ, that is, if $A^\star(\phi) = \mathrm{T}$ for every $\phi \in \Gamma$.

(b) We write

$$(3.66) \qquad\qquad \Gamma \models_\sigma \psi$$

to mean that for every σ-structure A, if A is a model of Γ then A is a model of ψ. The expression (3.66) is called a *semantic sequent*.

(c) We write

$$\Gamma \nvDash_\sigma \psi$$

to mean that (3.66) is not true.

When the context allows, we will ignore the subscript and write $\Gamma \models \psi$ instead of $\Gamma \models_\sigma \psi$. Using the Principle of Irrelevance, one can show that the choice of σ makes no difference as long as it contains all the propositional symbols in Γ and ψ (cf. the proof of Exercise 3.5.4).

Our aim will be to establish that for all Γ and ψ,

$$(3.67) \qquad\qquad \Gamma \vdash_\sigma \psi \quad \Leftrightarrow \quad \Gamma \models_\sigma \psi.$$

The two directions in (3.67) say very different things.

Going from left to right, (3.67) says that if there is a derivation with undischarged assumptions in Γ and conclusion ψ, then every model of Γ is a model of ψ. What would it mean for this to fail? It would mean that there is such a derivation D, and there is also a structure A in which all the formulas in Γ are true but ψ is false. Hence we would have derived a formula that is false in A from formulas that are true in A. This would be a devastating breakdown of our rules of proof: they should never derive something false from something true. So the left-to-right direction in (3.67) is verifying that we did not make some dreadful mistake when we set up the rules of natural deduction. The second half of this section will be devoted to this verification.

The direction from right to left in (3.67) says that if the truth of Γ guarantees the truth of ψ, then our natural deduction rules allow us to derive ψ from Γ. That is to say we do not need any more natural deduction rules besides those that we already have. We return to this in the next section.

Both directions of (3.67) are best read as saying something about our system of natural deduction. There are other proof calculi, and they have their own versions of (3.67). We can write $\Gamma \vdash_C \psi$ to mean that ψ is derivable from Γ in the proof calculus C. Then

$$\Gamma \vdash_C \psi \;\Rightarrow\; \Gamma \models \psi$$

is called *Soundness* of the calculus C, and the converse

$$\Gamma \models \psi \;\Rightarrow\; \Gamma \vdash_C \psi$$

is called *Adequacy* of the calculus C. The two directions together are called *Completeness* of C.

The main other proof calculi in common use among mathematical logicians are Hilbert-style calculi, the sequent calculus and tableaux. In Hilbert-style calculi one begins with axioms and draws consequences until one reaches the conclusion; there are no rules for discharging assumptions. The sequent calculus derives sequents from sequents rather than formulas from formulas. Tableau proofs prove ψ from Γ by trying systematically to describe a model of Γ that is not a model of ψ, and showing that there is no such model. Some other styles of proof calculus are more suitable for machine use.

Theorem 3.9.2 (Soundness of Natural Deduction for Propositional Logic) *Let σ be a signature, Γ a set of formulas of $LP(\sigma)$ and ψ a formula of $LP(\sigma)$. If $\Gamma \vdash_\sigma \psi$ then $\Gamma \models_\sigma \psi$.*

Proof The theorem states that

(3.68) If D is a σ-derivation whose conclusion is ψ and whose undischarged assumptions all lie in Γ, then every σ-structure that is a model of Γ is also a model of ψ.

Recall from Section 3.4 that D is a tree. We prove (3.68) for all Γ, ψ and D by induction on the height of this tree, as given by Definition 3.2.2(c). (Although we have turned the trees upside down, the definition of height is unchanged.)
Case 1: D has height 0. Then D is the derivation

$$\psi$$

which has ψ as both conclusion and undischarged assumption. So $\psi \in \Gamma$, and any model of Γ must be a model of ψ in particular.
Case 2: D has height $k \geqslant 1$, where we assume that (3.68) is true for all derivations D with height $< k$. Let R be the right label on the bottom node of D, which indicates what natural deduction rule was used to derive ψ. We divide

into subcases according to R. We consider only some typical cases, leaving the rest to the reader.

Case 2 (a): R is ($\rightarrow$I), so that by Definition 3.4.1(d)(i), D has the form

$$\frac{\displaystyle\frac{\not\phi}{\displaystyle\frac{D'}{\chi}}}{(\phi \rightarrow \chi)} \quad (\rightarrow\text{I})$$

where D' is a derivation of height $< k$ with conclusion χ. By Definition 3.4.1(g) the only formulas which can owe their dandahs to the bottom node of D are occurrences of ϕ (this is the meaning of the discharged ϕ written above at the top of D). So the undischarged assumptions of D' all lie in $\Gamma \cup \{\phi\}$. Let A be any σ-structure that is a model of Γ; we must show that it is also a model of $(\phi \rightarrow \chi)$. If it is not, then by the truth table for $\rightarrow$, we have $A^\star(\phi) = \text{T}$ and $A^\star(\chi) = \text{F}$. But this is impossible, since A is now a model of $\Gamma \cup \{\phi\}$ in which χ is false, and the induction assumption on D' says that there is no such model.

Case 2 (b): R is ($\rightarrow$E), so that by Definition 3.4.1(e)(i), D has the form

$$\frac{\begin{array}{cc} D_1 & D_2 \\ \phi & (\phi \rightarrow \psi) \end{array}}{\psi} \quad (\rightarrow\text{E})$$

where D_1 is a derivation of ϕ and D_2 a derivation of $(\phi \rightarrow \psi)$, both with their undischarged assumptions in Γ (since by (g), ($\rightarrow$E) authorises no dandahs). Both D_1 and D_2 have height $< k$. Let A be any σ-structure that is a model of Γ. We must show that it is also a model of ψ. Now the induction assumption on D_1 and D_2 implies that A is a model of both ϕ and $(\phi \rightarrow \psi)$. It follows by truth tables that A is a model of ψ as required.

Case 2 (c): R is (RAA), so that by Definition 3.4.1(d)(ii), D has the form

$$\frac{\displaystyle\frac{(\neg\psi)}{\displaystyle\frac{D'}{\bot}}}{\psi} \quad (\text{RAA})$$

where D' is a derivation of height $< k$ with conclusion $\bot$. By (g) the only formula that can be discharged thanks to the bottom node of D is $(\neg\psi)$, as indicated in the diagram. So the undischarged assumptions of D' all lie in $\Gamma \cup \{(\neg\psi)\}$. Let A be any σ-structure that is a model of Γ; we must prove that A is a model of ψ. We know that A is not a model of $\bot$, since no structure is a model of $\bot$. So by the induction assumption on D', A cannot be a model of $\Gamma \cup \{(\neg\psi)\}$. But A is a model of Γ by assumption; hence it cannot be a model of $\neg\psi$—in other words, it must be a model of ψ. ☐

Exercises

3.9.1. In the proof of Theorem 3.9.2, add clauses dealing with the cases where R is $(\wedge I)$ and where R is $(\vee E)$.

3.9.2 The following argument shows that Peirce's Formula $(((p \to q) \to p) \to p)$ (from Example 3.4.5) can't be proved using just the Axiom Rule and the rules $(\to I)$ and $(\to E)$; your task is to fill in the details. Instead of two truth values T and F, we introduce three truth values $1, \frac{1}{2}, 0$. Intuitively 1 is truth, 0 is falsehood and $\frac{1}{2}$ is somewhere betwixt and between. If ϕ has the value i and ψ has the value j (so $i, j \in \{1, \frac{1}{2}, 0\}$), then the value of $(\phi \to \psi)$ is

(3.69)   the greatest real number $r \leqslant 1$ such that $\min\{r, i\} \leqslant j$

(a) Write out the truth table for $\to$ using the three new truth values. (For example, you can check that $(p \to q)$ has the value $\frac{1}{2}$ when p has value 1 and q has value $\frac{1}{2}$.)

(b) Find values of p and q for which the value of $(((p \to q) \to p) \to p)$ is not 1.

(c) Show that the following holds for every derivation D using at most the Axiom Rule, $(\to I)$ and $(\to E)$: If ψ is the conclusion of D and A is a structure with $A^\star(\psi) < 1$, then $A^\star(\phi) \leqslant A^\star(\psi)$ for some undischarged assumption ϕ of D. [The argument is very similar to the proof of Theorem 3.9.2, using induction on the height of D.]

3.10 Completeness for propositional logic

Our target in this section is the following theorem. Throughout the section, σ is assumed to be the default signature $\{p_0, p_1, \dots\}$. We will briefly discuss this assumption at the end of the section.

Theorem 3.10.1 (Adequacy of Natural Deduction for Propositional Logic)
Let Γ be a set of formulas of $LP(\sigma)$ and ψ a formula of $LP(\sigma)$. If $\Gamma \models_\sigma \psi$ then $\Gamma \vdash_\sigma \psi$.

For simplicity the proof will use a stripped-down version of LP in which the only truth function symbols are

$$\wedge \quad \neg \quad \bot$$

Adding the other truth function symbols of LP makes the proof a bit longer but does not add any substantial new ideas. In any case we know from Corollary 3.8.9 that every formula of $LP(\sigma)$ is logically equivalent to one in which the only truth function symbols are $\wedge$, $\neg$ and $\bot$.

The proof will go in three steps, which we label as lemmas.

Definition 3.10.2 We say that a set Γ of formulas of $LP(\sigma)$ is *syntactically consistent* if $\Gamma \nvdash_\sigma \bot$. (This notion is independent of what signature σ we choose, so long as $LP(\sigma)$ contains all of Γ; cf. Exercises 3.4.3, 3.4.4.)

Lemma 3.10.3 *To prove the Adequacy Theorem it is enough to show that every syntactically consistent set of formulas of LP(σ) has a model.*

Proof of lemma Suppose every syntactically consistent set of formulas has a model. Let Γ be a set of formulas of $LP(\sigma)$ and ψ a formula of $LP(\sigma)$, and assume

$$(3.70) \qquad\qquad\qquad \Gamma \models_\sigma \psi$$

Then we claim that

$$(3.71) \qquad\qquad \Gamma \cup \{(\neg\psi)\} \text{ has no models}$$

For by (3.70) every model of Γ is a model of ψ, and so not a model of $(\neg\psi)$.

Since we assume that every syntactically consistent set has a model, it follows from (3.71) that $\Gamma \cup \{(\neg\psi)\}$ is not syntactically consistent, that is,

$$(3.72) \qquad\qquad \Gamma \cup \{(\neg\psi)\} \vdash_\sigma \bot$$

Then the correctness of the sequent $\Gamma \vdash_\sigma \psi$ follows by Example 3.4.3. $\square$

Jaakko Hintikka Finland and USA, living.
How to construct a set of formulas that gives an exact description of a situation.

The remaining steps of the proof rest on the following definition.

Definition 3.10.4 We say that a set Γ of formulas of (the stripped-down) LP is a *Hintikka set* (for LP) if it has the following properties:

(1) If a formula $(\phi \wedge \psi)$ is in Γ then ϕ is in Γ and ψ is in Γ.

(2) If a formula $(\neg(\phi \wedge \psi))$ is in Γ then at least one of $(\neg\phi)$ and $(\neg\psi)$ is in Γ.

(3) If a formula $(\neg(\neg\phi))$ is in Γ then ϕ is in Γ.

(4) $\bot$ is not in Γ.

(5) There is no propositional symbol p such that both p and $(\neg p)$ are in Γ.

Lemma 3.10.5 *Every Hintikka set has a model.*

Proof of lemma Let Γ be a Hintikka set whose formulas are in $\mathrm{LP}(\sigma)$. Let A be the following σ-structure:

(3.73)
$$A(p) = \begin{cases} \mathrm{T} & \text{if } p \in \Gamma \\ \mathrm{F} & \text{otherwise} \end{cases}$$

We will show that for every formula ϕ of $\mathrm{LP}(\sigma)$, both (a) and (b) are true:

(3.74)
(a) If ϕ is in Γ then $A^\star(\phi) = \mathrm{T}$
(b) If $(\neg\phi)$ is in Γ then $A^\star(\phi) = \mathrm{F}$

We prove this by induction on the complexity of ϕ, using Definition 3.5.6.

Case 1: ϕ is a propositional symbol p. Then $A^\star(p) = A(p)$. If $\phi \in \Gamma$ then $A^\star(p) = \mathrm{T}$ by (3.73), proving (a). If $(\neg p) \in \Gamma$ then by property (5) of Hintikka sets, $p \notin \Gamma$, so $A^\star(p) = \mathrm{F}$ by (3.73), proving (b).

Case 2: ϕ is $\bot$. Then by property (4) of Hintikka sets, $\bot \notin \Gamma$, so (a) of (3.74) holds trivially. Also by truth tables $A^\star(\bot) = \mathrm{F}$ so that (b) holds too.

Case 3: ϕ is $(\neg\psi)$ for some proposition ψ; by induction assumption (3.74) holds for ψ. If $\phi \in \Gamma$ then $(\neg\psi) \in \Gamma$, so $A^\star(\psi) = \mathrm{F}$ by (b) for ψ, and hence $A^\star(\phi) = \mathrm{T}$. This proves (a) for ϕ. For (b), suppose $(\neg\phi) \in \Gamma$. Then $(\neg(\neg\psi)) \in \Gamma$, and so by property (3) of Hintikka sets, $\psi \in \Gamma$, so that $A^\star(\psi) = \mathrm{T}$ by (a) for ψ. But then $A^\star(\phi) = A^\star((\neg\psi)) = \mathrm{F}$, proving (b) for ϕ.

Case 4: ϕ is $(\psi \wedge \chi)$; by induction assumption (3.74) holds for both ψ and χ. To prove (a) for ϕ, if $\phi \in \Gamma$ then by property (1) of Hintikka sets, $\psi \in \Gamma$ and $\chi \in \Gamma$. So by induction assumption (a), $A^\star(\psi) = A^\star(\chi) = \mathrm{T}$. But then $A^\star((\phi \wedge \psi)) = \mathrm{T}$, proving (a) for ϕ. For (b), suppose $(\neg\phi) \in \Gamma$. Then by property (2) of Hintikka sets, at least one of $(\neg\psi)$ and $(\neg\chi)$ is in Γ; say $(\neg\psi) \in \Gamma$. Then by induction assumption (b) for ψ, $A^\star(\psi) = \mathrm{F}$. It follows that $A^\star(\phi) = \mathrm{F}$, proving (b) for ϕ.

This proves (3.74) for every formula ϕ of $\mathrm{LP}(\sigma)$. By (a) of (3.74), A is a model of Γ. $\qquad\square$

Lemma 3.10.6 *If Γ is a syntactically consistent set of formulas of $LP(\sigma)$, then there is a Hintikka set Δ of formulas of $LP(\sigma)$ with $\Gamma \subseteq \Delta$.*

Proof of lemma   This is quite a busy proof, because we have to work to build Δ. We start by assuming that Γ is syntactically consistent.

In Exercise 3.2.5 we saw how to assign to each formula ϕ of $LP(\sigma)$ a distinct positive integer $GN(\phi)$, called its *Gödel number*. (This is the only place in this proof where we use the assumption that σ is the set $\{p_0, p_1, \dots\}$.) So we can list all the formulas of $LP(\sigma)$ in increasing order of their Gödel numbers, say as

$$\psi_0, \psi_1, \psi_2, \dots$$

We need a slightly different listing of the formulas, got by taking them as follows:

$$\psi_0,$$
$$\psi_0, \psi_1,$$
$$\psi_0, \psi_1, \psi_2,$$
$$\psi_0, \psi_1, \psi_2, \psi_3,$$
$$\dots$$

Write

$$\phi_0, \phi_1, \phi_2, \phi_3, \dots$$

for this listing of the formulas of $LP(\sigma)$. It has the property that every formula appears infinitely often in the list.

We shall build a sequence of sets $\Gamma_0 \subseteq \Gamma_1 \subseteq \cdots$, one by one. The idea of the construction is to take the requirements for a Hintikka set, one by one; as we come to each requirement, we build the next Γ_i so that the requirement is met.

We put $\Gamma_0 = \Gamma$. Then when Γ_i has just been defined, we define Γ_{i+1} as follows, depending on the formula ϕ_i.

(α) If ϕ_i is not in Γ_i, or is $\bot$ or $\neg\bot$ or either p or $(\neg p)$ for some propositional symbol p, we put $\Gamma_{i+1} = \Gamma_i$. (The reason is that in these cases the definition of a Hintikka set does not tell us to do anything with ϕ_i if we want to make Γ_i into a Hintikka set.)

(β) If ϕ_i is in Γ_i and is $(\chi_1 \wedge \chi_2)$, then we put $\Gamma_{i+1} = \Gamma_i \cup \{\chi_1, \chi_2\}$. (By (1) in the definition of Hintikka sets, any Hintikka set containing Γ_i would have to contain these two formulas.)

(γ) If ϕ_i is in Γ_i and is $(\neg(\chi_1 \wedge \chi_2))$, and $\Gamma_i \cup \{\neg\chi_1\}$ is syntactically consistent, then we put $\Gamma_{i+1} = \Gamma_i \cup \{\neg\chi_1\}$; if it is not syntactically consistent, then instead we put $\Gamma_{i+1} = \Gamma_i \cup \{\neg\chi_2\}$. (A similar reason applies, using (2) in the definition of Hintikka sets.)

(δ) If ϕ_i is in Γ_i and is $(\neg(\neg\phi))$, we put $\Gamma_{i+1} = \Gamma_i \cup \{\phi\}$. (Similarly, using (3) in the definition of Hintikka sets.)

We take Δ to be the union $\Gamma_0 \cup \Gamma_1 \cup \cdots$ (in other words, the set of all things that are members of Γ_n for at least one natural number n). We claim:

(3.75) Δ is syntactically consistent

If this claim is false, then there is some derivation D of $\bot$ whose undischarged assumptions are in Δ. Since derivations contain only finitely many symbols, there are finitely many undischarged assumptions in D, and so they must all be already in some Γ_i. So the claim follows if we prove, by induction on i, that each Γ_i is syntactically consistent.

Case 1: $i = 0$. We chose Γ_0 to be Γ, which was assumed to be syntactically consistent.

Case 2: $i = k + 1$, assuming Γ_k is syntactically consistent. Here the argument depends on how Γ_{k+1} was constructed.

Suppose first that $\Gamma_{k+1} = \Gamma_k$. Then by induction hypothesis Γ_{k+1} is syntactically consistent.

Next suppose Γ_{k+1} was made by adding one or two formulas to Γ_k, depending on ϕ_k. There are three separate cases for the possible forms of ϕ_k.

(i) Suppose ϕ_k is in Γ_k and is $(\chi_1 \wedge \chi_2)$. This was case (β) in the construction of Γ_{k+1}, so that $\Gamma_{k+1} = \Gamma_k \cup \{\chi_1, \chi_2\}$. Assume for contradiction that there is a derivation D of $\bot$ from Γ_{k+1}. Then at each place in D where χ_1 occurs as an undischarged assumption of D, replace χ_1 by

$$\frac{(\chi_1 \wedge \chi_2)}{\chi_1} \ (\wedge E)$$

and similarly with χ_2. The result is a derivation of $\bot$ whose undischarged assumptions all lie in Γ_k. This contradicts the induction hypothesis that Γ_k was syntactically consistent.

(ii) Suppose ϕ_k is in Γ_k and is $(\neg(\chi_1 \wedge \chi_2))$. This was case ($\gamma$) above. Assume for contradiction that there is a derivation D whose conclusion is $\bot$ and whose undischarged assumptions all lie in Γ_{k+1}. In the construction of Γ_{k+1} at (γ), we took Γ_{k+1} to be $\Gamma_k \cup \{(\neg\chi_1)\}$ if this set was syntactically consistent; since Γ_{k+1} is not syntactically consistent, we infer that

$\Gamma_k \cup \{(\neg\chi_1)\}$ is not syntactically consistent.

Hence by (γ), Γ_{k+1} is $\Gamma_k \cup \{(\neg\chi_2)\}$. Therefore,

$\Gamma_k \cup \{(\neg\chi_2)\}$ is not syntactically consistent.

So for $j = 1, 2$ there are derivations D_j of $\bot$ whose undischarged assumptions are in $\Gamma_k \cup \{(\neg\chi_j)\}$. Then we have the following derivation

$$
\cfrac{
  \cfrac{
    \cfrac{
      \cfrac{\cancel{(\neg\chi_1)}}{\quad D_1 \quad}}{\bot}\ (\text{RAA})
  }{\chi_1}
  \qquad
  \cfrac{
    \cfrac{
      \cfrac{\cancel{(\neg\chi_2)}}{\quad D_2 \quad}}{\bot}\ (\text{RAA})
  }{\chi_2}
}{(\chi_1 \wedge \chi_2)}\ (\wedge\text{I})
\qquad (\neg(\chi_1 \wedge \chi_2))
$$
$$
\bot \qquad (\neg\text{E})
$$

which proves $\Gamma_k \vdash_\sigma \bot$ and so contradicts the induction hypothesis that Γ_k was syntactically consistent.

(iii)   Suppose ϕ_k is in Γ_k and is $(\neg(\neg\phi))$. This was case (δ), so that $\Gamma_{k+1} = \Gamma_k \cup \{\phi\}$. Assume for contradiction that there is a derivation D of $\bot$ whose undischarged assumptions are in Γ_{k+1}. By adding a derivation of ϕ from $(\neg(\neg\phi))$ (cf. Example 2.6.3) at every undischarged occurrence of the assumption ϕ in D, we turn D into a derivation of $\bot$ from Γ_k. Again this contradicts the induction hypothesis that Γ_k is syntactically consistent.

Claim (3.75) is proved.

Since $\Gamma_0 = \Gamma$, we have $\Gamma \subseteq \Delta$. We check that Δ meets all the conditions for a Hintikka set.

Condition (1) says that if $(\phi \wedge \psi)$ is in Δ then ϕ and ψ are also in Δ. Suppose then that $(\phi \wedge \psi)$ is in Δ; then by construction of Δ it must already be in some Γ_i. Now in the listing of formulas the formula $(\phi \wedge \psi)$ appears as ϕ_j for some $j \geqslant i$. We consider what happens in the construction of Γ_{j+1}. By case (β) in the construction, both ϕ and ψ are in Γ_{j+1}, so they are in Δ as required.

Condition (2) says that if $(\neg(\phi \wedge \psi))$ is in Δ then at least one of $(\neg\phi)$ and $(\neg\psi)$ is in Δ. We prove this in the same way as condition (1), but using (γ) in the construction of Δ. Likewise, condition (3) is proved using (δ).

Condition (4) says that $\bot$ is not in Δ. This follows from the fact that Δ is syntactically consistent, which was the claim (3.75). Finally condition (5) says that there is no propositional symbol p for which both p and $(\neg p)$ are in Δ. But if there were, then $(\neg\text{E})$ would immediately give us a derivation of $\bot$ from assumptions in Δ, contradicting claim (3.75). $\square$

Proof of the Adequacy Theorem By Lemma 3.10.3 it is enough to show that every syntactically consistent set of formulas of LP(σ) has a model. Suppose Γ is a syntactically consistent set of formulas of LP(σ). By Lemma 3.10.6 there is a Hintikka set Δ of formulas of LP(σ) with $\Gamma \subseteq \Delta$. By Lemma 3.10.5, Δ has a model A. Since every formula in Γ is also in Δ, A is a model of Γ too, as required. $\square$

Theorem 3.10.7 (Completeness Theorem) *Let Γ be a set of formulas of LP(σ) and ψ a formula of LP(σ). Then*

$$\Gamma \vdash_\sigma \psi \;\Leftrightarrow\; \Gamma \models_\sigma \psi.$$

Proof This is the Soundness and Adequacy Theorems combined. $\square$

We should comment on the assumption that σ is the default signature $\{p_0, p_1, \dots\}$. If you have a signature τ that is different from this, then probably the 'symbols' in τ are literally symbols that can be written on a page. So with a little ingenuity it must be possible to list them, say as

$$q_0, q_1, \dots$$

If there are infinitely many, then our proof of the Adequacy Theorem works with q_i's in place of the corresponding p_i's. If τ has only finitely many symbols, the proof still works, since the assumption that there are infinitely many p_i's was never used.

Problems arise only if your signature has more abstract 'symbols'. For example, if you want to regard each real number as a symbol in σ, then it follows from Cantor's Theorem, Corollary 7.8.6, that the formulas cannot be listed in the way our proof assumed. Apart from the problem of writing formulas down, the theorems of this chapter do still hold in this generalised setting. But they need generalised proofs that employ more set theory than we want to assume in this book.

Exercises

3.10.1. If we kept the truth function symbols $\wedge$, $\vee$ and $\leftrightarrow$ in the language, we would need

    (a) clauses for them in the Definition 3.10.4 of Hintikka sets,

    (b) clauses for them in the proof of Lemma 3.10.5 and

    (c) clauses for them in the proof of Lemma 3.10.6.

Write the required clauses, using the examples in the text as a guide.

3.10.2. Let σ and τ be signatures. Suppose Φ is a set of formulas of LP(σ) and Ψ is a set of formulas of LP(τ). We say that (Φ, Ψ) is a *Craig pair* if there is no formula θ of LP($\sigma \cap \tau$) such that $\Phi \vdash_\sigma \theta$ and $\Psi \vdash_\tau \neg\theta$.

    (a) Show that if (Φ, Ψ) is a Craig pair then both Φ and Ψ are syntactically consistent.

(b)   Show that if (Φ, Ψ) is a Craig pair and both Φ and Ψ are Hintikka sets of formulas of $LP(\sigma)$ and $LP(\tau)$, respectively, then $\Phi \cup \Psi$ is a Hintikka set.

(c)   Show that if (Φ, Ψ) is a Craig pair, then there are Hintikka sets Φ' of formulas of $LP(\sigma)$, and Ψ' of formulas of $LP(\tau)$, such that $\Phi \subseteq \Phi'$ and $\Psi \subseteq \Psi'$ and (Φ', Ψ') is a Craig pair. [The proof is very similar to that of Lemma 3.10.6. List all the formulas of $LP(\sigma \cup \tau)$.]

(d)   Deduce from (a)–(c) that if ϕ is a formula of $LP(\sigma)$, ψ is a formula of $LP(\tau)$ and $\{\phi\} \models_{\sigma \cup \tau} \psi$, then there is a formula θ of $LP(\sigma \cap \tau)$, such that $\{\phi\} \vdash_\sigma \theta$ and $\{\theta\} \vdash_\tau \psi$. (This result is known as *Craig's Interpolation Theorem*; the formula θ is the *interpolant*. Craig's Interpolation Theorem can be extended to first-order logic, by a very similar proof to that in this exercise.)

4   First interlude: Wason's selection task

You will be shown four cards. Each of these cards has a number written on one side and a letter on the other. You will also be shown a statement S about the cards. Then you must answer the following question:

Which card or cards must I turn over in order to check whether the statement S is true?

Here is the statement S:

S: If a card has a vowel on one side,
it has an even number on the other side.

Here are the cards:

E   K   4   7

Write down your answer to the question before continuing.

Logical analysis

The following table shows when the statement 'If the card has a vowel on one side then it has an even number on the other side' is true.

Vowel	Even	(Vowel → Even)
T	T	T
T	F	F
F	T	T
F	F	T

What do we need to check in order to be sure that the statement S is true for each card? In the table, we must check that no card is in the second row, that is,

Every card has either a consonant or an even number (or both).

With this in mind:

Card	Verdict
E	No consonant or even number visible: MUST CHECK
K	Has consonant, NEED NOT CHECK FURTHER
4	Has even number, NEED NOT CHECK FURTHER
7	No consonant or even number visible: MUST CHECK

So the cards to check are E and 7.

Psychological analysis

When we gave this question to our logic class, after they had studied Chapter 3, they gave the following answers:

E and 7 (right answer)	0%
E	50%
E and 4	20%
K and 7	15%
7	5%
K	5%
All cards	5%

These results are fairly typical. There are several ways of adjusting the experiment, and a good deal is known about which adjustments make it easier or harder to reach the 'logically correct' answer.

One possible explanation is that human beings do not have any built-in understanding of truth tables. We can make quick decisions on the basis of rules that evolution built into us, and for some reason these rules tell us to choose card E but not card 7 in this experiment. Psychologists have made various suggestions of why this should be. An explanation should also explain why certain versions of the experiment make the task easier and some make it harder. See, for example, Keith E. Stanovich, *Who is Rational? Studies of Individual Differences in Reasoning*, Lawrence Erlbaum, Mahwah, NJ, 1999, or Keith Stenning, *Seeing Reason: Image and Language in Learning to Think*, Oxford University Press, Oxford, 2002.

This experiment is known as the *Wason Selection Task*. It was first devised by Peter Wason in 1966, and since then it has been one of the most intensely researched experiments in cognitive science.

Of course, the psychologists' results do not mean we cannot do truth tables. But they do seem to show that in order to use truth tables reliably, we need to take enough time to train ourselves, and then to make the appropriate calculations, preferably on paper.

5 Quantifier-free logic

Here are two simple-minded arguments that we might meet any day in a piece of mathematics:

(a) $x = y$, so $\cos x = \cos y$.

(b) All positive real numbers have real square roots, and π is a positive real number, so π has a real square root.

Though the arguments are simple, they are beyond the capacities of Chapters 2 and 3. In the present chapter we will formalise (a), and Chapter 7 will take care of (b).

Our formalisation will use a language LR, the 'Language of Relations'. Like LP, it allows different choices of signature. The languages LR(σ), for any signature σ, are known as *first-order languages*. (The name contrasts them with *second-order languages*; see (7.43).) The proof rules and semantics of first-order languages are known as *first-order logic* or as *predicate logic*.

5.1 Terms

A *term* is an expression that can be used for naming something. (Well, *any* expression can be used for naming your cat. But we are talking about names within the usual conventions of mathematics.)

The most common examples of terms are constant symbols, variables, definite descriptions and complex mathematical terms.

(a) **Constant symbols:** These are single mathematical symbols that are used as fixed names of particular numbers, functions or sets. For example,

$$0, \ 1, \ 42, \ \pi, \ \infty, \ \mathbb{R}, \ \emptyset, \ \cos.$$

(b) **Variables:** These are letters that are used to stand for any numbers or any elements of some particular set. For example,

$$x, \ y, \ z, \ \text{etc.} \qquad \text{ranging over real numbers,}$$
$$m, \ n, \ p, \ \text{etc.} \qquad \text{ranging over integers,}$$
$$f, \ g, \ F, \ G, \ \text{etc.} \quad \text{ranging over functions of some kind.}$$

(c) **Definite descriptions:** These are singular expressions of the form 'the ...' or 'this ...' or 'our ...', etc. For example,

> the number,
> the number $\sqrt{\pi}$,
> the last nonzero remainder in the above steps,
> this graph G,
> our original assumption,
> its top node,
> Proposition 2.6.

(Think of the last as short for 'the proposition numbered 2.6'.)

(d) **Complex mathematical terms:** These are mathematical expressions that do the same job as definite descriptions, but in purely mathematical notation. For example,

$$\sqrt[3]{x + y}$$

is a complex mathematical term; it means the same as 'the cube root of the result of adding x to y'.

Also,

$$\int_{-\pi}^{\pi} z^2 \, dz$$

is a complex mathematical term meaning the same as 'the integral from $-\pi$ to π of the function $(z \mapsto z^2)$'.

Free and bound variables

Consider the integral

(5.1)
$$\int_{x}^{y} z^2 \, dz$$

There are three variables in (5.1): x, y, z. The variable z occurs twice.

There is an important difference between x and y on the one hand, and z on the other. *We can meaningfully read x and y as naming particular numbers.* For example, the integral still makes sense if we have already put $x = 1$ and $y = 2$. We can check this by replacing x and y:

$$\int_{1}^{2} z^2 \, dz$$

We cannot do the same with z. For example, this is nonsense:

$$\int_{1}^{2} \pi^2 \, d\pi$$

To make sense of it, we would have to forget that π means a certain number between 3 and 4. (These rewritings are examples of the *rewriting test* for freedom and boundness.)

Definition 5.1.1 An occurrence of a variable in a piece of mathematical text is said to be *free* if the text allows us to read the variable as a name. If not free, the occurrence is *bound*. When an occurrence of a variable in the text is bound, there must always be something in the text that prevents us taking the occurrence as a name. For example, the integral sign in (5.1) causes both occurrences of z to be bound; we say that the integral *binds* the occurrences.

The talk of occurrences in Definition 5.1.1 is necessary, because it can happen that the same variable has free occurrences and bound occurrences in the same text. For example:

$$3z + \int_1^2 z^2 \, dz$$

Here the first z is free but the other two occurrences of z are bound by the integral.

It is also necessary to mention the surrounding text. An occurrence of a variable can be free in text T_1 but bound in text T_2 that contains T_1. For example, the integral (5.1) contains the expression

(5.2) $$z^2$$

In (5.2), z is free, although it is bound in (5.1). This is a very common situation: probably most occurrences of variables become bound if we take a large enough piece of surrounding text.

The rewriting test must be used with care. For example, it makes sense to write

$$\int_x^y \pi^2 \, dz$$

which suggests that the first occurrence of z in (5.1) was really free. But this is wrong. There is no way that we can read the first z in the integral (5.1) as standing for π and the second z as the variable of integration. As mathematicians say, the two occurrences are 'the same z'. In order to make Definition 5.1.1 precise, we would need to restrict it to a precisely defined language. In Section 7.2 we will replace it by a precise formal definition for the language $\mathrm{LR}(\sigma)$.

Integrals are not the only expressions that bind variables. For example,

(5.3) $$\{c \mid |c| > 2\}$$

Here both occurrences of c are bound by the set expression.

(5.4)   For every integer n there is an integer $m > n$

In this example the variable n is bound by the expression 'For every integer'; it does not make sense to say 'For every integer 13.5', for example. Also the variable m is bound by the expression 'there is an integer'.

(5.5)   For every $\epsilon > 0$ there is a $\delta > 0$ such that for every
                 real x with $|x| < \delta$ we have $f(x) < \epsilon$

Here the variable ϵ is bound by 'For every' and the variable δ is bound by 'there is'. The symbol f is free in the sentence—for example, it could stand for the sine function.

More examples of free and bound occurrences are in the exercises.

The language $\mathrm{LR}(\sigma)$ will have symbols to stand as variables. It will also have a symbol $\forall$ to stand for the expression 'for every', and a symbol $\exists$ for the expression 'there is'. (You may have noticed that $\forall$ and $\exists$ come from the capitalised letters in 'for All' and 'there Exists', rotated through angle π.) The symbol $\forall$ is known as the *universal quantifier symbol*, and the symbol $\exists$ as the *existential quantifier symbol*. These two quantifier symbols will be part of the lexicon of the language LR, and in fact they will be the only symbols in LR that bind variables.

In Chapter 7 we will see how to interpret the quantifier symbols in the semantics of LR, and we will give a precise definition of 'free' and 'bound' occurrences of variables in formulas of LR. But for later sections of the present chapter we will confine ourselves to formulas of LR in which no quantifier symbols occur, that is, to *quantifier-free* (qf) formulas. In these formulas there are no bound variables.

Exercises

5.1.1.   Find all the terms in the following paragraph:

"Putting $\theta = \frac{\pi}{9}$ and $c = \cos\frac{\pi}{9}$, Example 6.3 gives

$$\cos 3\theta = 4c^3 - 3c.$$

However, $\cos 3\theta = \cos\frac{\pi}{3} = \frac{1}{2}$. Hence $\frac{1}{2} = 4c^3 - 3c$. In other words, $c = \cos\frac{\pi}{9}$ is a root of the cubic equation

$$8x^3 - 6x - 1 = 0.\text{"}$$

5.1.2. Which occurrences of variables are free and which are bound in the following statements? Where an occurrence of a variable is bound, explain what binds it.

(a) $v = 3$.

(b) $\lim\limits_{t \to 0} \dfrac{\sin t + x}{t} = y$.

(c) For all real r, $e^r > r$.

(d) For some real r, $e^r > r$.

(e) $z\dfrac{d^2 y}{dx^2} = x^2 + z + 1$.

5.1.3. Find 10 occurrences of terms in the following text. For each one, say whether it is (a) a constant symbol, (b) a variable, (c) a definite description, (d) a complex mathematical term. If it is a variable, say whether it is free or bound in the text.

"Since

$$f(x) = x^3 - x^2 - 2x = x(x+1)(x-2),$$

the zeros of the function f are 0, −1 and 2. The zeros partition the interval $[-1, 2]$ into two subintervals: $[-1, 0]$ and $[0, 2]$. We integrate f over each subinterval. The integral over $[-1, 0]$ is

$$\int_{-1}^{0} (x^3 - x^2 - 2x)dx = \left[\frac{x^4}{4} - \frac{x^3}{3} - x^2 \right]_{-1}^{0} = \frac{5}{12}."$$

5.1.4. For each of the following expressions, write a true mathematical sentence containing the expression, so that all the variables in the expression are bound in your sentence.

(a) $x < y + z$.

(b) $|u - v|$.

(c) $\cos^n \theta$.

[The variables in (c) are n and θ.]

5.2 Relations and functions

Definition 5.2.1 The *arity* (also called *rank*) of a statement or a term is the number of variables that have free occurrences in it. (If a variable has two or more free occurrences, it still counts for just one in the arity.) We say that a statement or term of arity n is *n-ary*; *binary* means 2-ary. A statement of arity ≥ 1 is called a *predicate*.

Example 5.2.2 The statement

$$\text{If } p > -1 \text{ then } (1 + p)^n \geqslant 1 + np$$

is a binary predicate, with free variables p and n. The term

$$\int_x^y \tan z \; dz$$

is also binary, with free variables x and y.

Definition 5.2.3 Suppose X is a set and n is a positive integer. Then an *n-ary relation on* X is a set of ordered n-tuples of elements of X. (And henceforth we abbreviate 'ordered n-tuple' to 'n-tuple'.)

There are two main ways of naming a relation. The first is to list the n-tuples that are in it. For example, here is a binary relation on the set $\{0, 1, 2, 3, 4, 5\}$:

$$(5.6) \qquad \begin{aligned} &\{(0,1), (0,2), (0,3), (0,4), (0,5), (1,2), (1,3), (1,4), \\ &(1,5), (2,3), (2,4), (2,5), (3,4), (3,5), (4,5)\} \end{aligned}$$

This is the 'less than' relation on $\{0, 1, 2, 3, 4, 5\}$.

This method is of no use when the relation has too many n-tuples to list. In such cases, we name the relation by giving an n-ary predicate. We need a convention for fixing the order of the variables. For example, if we take the predicate

$$y \geqslant x$$

and ask what relation it defines on $\{0, 1, 2, 3\}$, does x go before y (so that $(0, 1)$ is in the relation) or the other way round (so that $(1, 0)$ is in the relation)? One standard solution to this problem is to give the definition of the relation as follows:

$$(5.7) \qquad\qquad\qquad R(x, y) \;\Leftrightarrow\; y \geqslant x$$

The variables on the left are listed with x first, so we know that x corresponds to the first place in the ordered pairs. After this definition has been given, we can use R as a name of the relation defined.

More precisely, we test whether a pair $(3, 2)$ is in the relation R by putting 3 for x and 2 for y in the defining predicate:

$$(5.8) \qquad\qquad\qquad 2 \geqslant 3$$

This statement (5.8) is false, so the pair $(3, 2)$ is not in the relation R. But '$2 \geqslant 2$' is true, so $(2, 2)$ is in R. In view of this and similar examples, if R is a binary predicate, we often write 'xRy' instead of '$R(x, y)$'.

A definition of a relation in the form (5.7) has to meet certain conditions, or it will not work (see Exercise 5.2.3). First, we must be told what set the variables range over. (For (5.7) this set is $\{0, 1, 2, 3\}$.) Second, the variables listed inside parentheses after R must all be distinct from each other. And third, every free variable in the predicate on the right should appear among these variables listed inside the parentheses.

A particular kind of relation which will interest us is known as a (strict) linear order.

Definition 5.2.4 Suppose X is a set. A *strict linear order* on X is a binary relation R on X such that

(1) for all $x \in X$, xRx is false;

(2) for all x, y and $z \in X$, if xRy and yRz then xRz;

(3) for all $x, y \in X$, either xRy or yRx or $x = y$.

As an example, let X be a subset of $\mathbb{R}$, and define

$$xRy \iff x < y$$

on X (where $<$ is the usual linear order of the real numbers).

If R is a strict linear order on X, and Y is a subset of X, then the restriction of R to Y is a strict linear order on Y.

Definition 5.2.5 Suppose R is a strict linear order on a set X. An element $y \in X$ is called a *least element* of X if yRx whenever $x \in X$ and $x \neq y$. An element $z \in X$ is called a *greatest element* of X if xRz whenever $x \in X$ and $x \neq z$.

It is easy to see that there can be at most one least element and at most one greatest element in a strict linear order.

Example 5.2.6 We consider strict linear orders on a finite set. Suppose X is a finite non-empty set and R is a strict linear order on X. Then we claim that there is a least element in X. For if not, let $y_1 \in X$. Since y_1 is not least, there is an element $y_2 \in X$ such that y_2Ry_1, by (3) in Definition 5.2.4. Since y_2 is not least, we can choose $y_3 \in X$ such that y_3Ry_2. Continuing, we construct $y_4, y_5, \ldots$. By repeated use of (2), we see that y_iRy_j whenever $i > j$, so $y_i \neq y_j$ by (1). This contradicts the finiteness of X.

Now let x_1 be a least element of X. Then $X \setminus \{x_1\}$ (the set consisting of everything in X except x_1) is also finite, so if non-empty it has a least element x_2, and x_1Rx_2. Similarly $X \setminus \{x_1, x_2\}$, if non-empty, has a least element x_3 and x_2Rx_3. Since X is finite, this procedure must stop eventually to give $X = \{x_1, \ldots, x_n\}$ with

(5.9) $$x_1Rx_2, \ x_2Rx_3, \ldots, \ x_{n-1}Rx_n$$

By repeated use of (2), $x_i R x_j$ whenever $i < j$. So X can be listed in R-increasing order.

Conversely, if X is any finite set and we enumerate X without repetitions as $X = \{x_1, \ldots, x_n\}$, and define

$$x R y \iff x = x_i \quad \text{and} \quad y = x_j \text{ for some } i < j,$$

then we obtain a strict linear order R on X.

The symbol $<$ is often used for strict linear orders, even if they have nothing to do with the usual order on the real numbers. Correspondingly, when we come to define the signature of linear order in Example 5.3.2(c), we will put the symbol $<$ in its signature, to stand for any strict linear order relation. (Likewise if you have studied groups you will know that the symbol $\cdot$ in $x \cdot y$ refers to different functions in different groups.)

We have assumed familiarity with the idea of a function. A function can actually be viewed as a special kind of relation, as follows.

Definition 5.2.7 Suppose n is a positive integer, X is a set and R is an $(n+1)$-ary relation on X. We say that R is an *n-ary function* on X if the following holds:

(5.10)   For all $a_1, \ldots, a_n$ in X there is exactly one b in X such that $(a_1, \ldots, a_n, b)$ is in R

When this condition holds, the unique b is called the *value* of the function at $(a_1, \ldots, a_n)$.

For example, the relation (5.6) is not a function on $\{0, 1, 2, 3, 4, 5\}$: there is no pair with 5 on the left, and there are five pairs with 0 on the left. But the following is a 1-ary function on $\{0, 1, 2, 3, 4, 5\}$:

(5.11) $\{(0, 1), (1, 2), (2, 3), (3, 4), (4, 5), (5, 0)\}$

This function is sometimes known as 'plus one mod 6'.

Since a function is a kind of relation, we can define particular functions in the same ways as we define particular relations. But there is another way that is particularly appropriate for functions. Instead of an $(n+1)$-ary predicate, we use an n-ary term, as in the following example:

(5.12) $F(x, y) = $ the remainder when $y - x$ is divided by 6

We use $=$ instead of $\iff$. But again for an n-ary function we have n variables listed on the left, and the order of the listing shows which places they belong to in the n-tuples. Also after this definition has been given, the function symbol F is available as a name of the defined function. For example, on $\{0, 1, 2, 3, 4, 5\}$, the remainder when $4 - 3$ is divided by 6 is 1, whereas the remainder when $3 - 4$ is divided by 6 is 5, so that

$$F(3, 4) = 1, \qquad F(4, 3) = 5.$$

Remark 5.2.8 *Overloading* means using an expression with two different arities. For example, in ordinary arithmetic we have

$$-6$$

with $-$ of arity 1, and

$$8 - 6$$

with $-$ of arity 2. So the symbol $-$ is overloaded. Likewise, some computer languages deliberately use overloading. But in this course overloading is likely to cause confusion, so we avoid it. A relation or function symbol has only one arity.

There is another kind of overloading. Compare

(a) $\cos(\pi) = -1$.

(b) $\cos$ is differentiable.

In the first case, cos is a 1-ary function symbol; in the second it is a term. The difference is between applying the function and talking about it. Again we will not allow this in our formal languages. In the terminology of the next section, cos in (a) is a function symbol and not a constant symbol, and vice versa in (b).

Substitutions

To test whether a pair (a, b) is in the relation R of (5.7), we substituted a name of a for x and a name of b for y in the predicate '$y \geqslant x$', and we asked whether the resulting sentence is true. Normally this method works. But here is a case where we get strange results by substituting a name for a variable in a predicate.

Consider the equation

(5.13)
$$\int_1^2 2(x + y) \, dx = 5$$

which is a 1-ary predicate with the variable y. To integrate (5.13), we regard y as constant and we get

$$5 = \left[x^2 + 2xy \right]_1^2 = (4 + 4y) - (1 + 2y) = 3 + 2y$$

which you can solve to get $y = 1$.

Now suppose we give x the value 1, and we try to say that 1 is a solution for y in equation (5.13), by writing x in place of y. We get

$$5 = \int_1^2 2(x+x)\,dx = \int_1^2 4x\,dx = \left[2x^2\right]_1^2 = 8 - 2 = 6.$$

So $5 = 6$. What went wrong?

Our mistake was that when we put x in place of the free occurrence of y in the equation, x *became bound* by the integration over x.

Definition 5.2.9 When a term t can be substituted for all free occurrences of a variable y in an expression E, without any of the variables in t becoming bound by other parts of E, we say that t is *substitutable for* y in E. (Some logicians say 'free for' rather than 'substitutable for'.)

The problem with our integral equation (5.13) was that the term x is not substitutable for y in the equation.

Definition 5.2.10 Suppose E is an expression, $y_1, \dots, y_n$ are distinct variables and $t_1, \dots, t_n$ are terms such that each t_i is substitutable for y_i in E. Then we write

(5.14) $E[t_1/y_1, \dots, t_n/y_n]$

for the expression got from E by simultaneously replacing each free occurrence of each y_i in E by t_i. If some t_i is not substitutable for y_i in E, we count the expression $E[t_1/y_1, \dots, t_n/y_n]$ as meaningless. The expression

$$t_1/y_1, \dots, t_n/y_n$$

in (5.14) is called a *substitution for variables*, or a *substitution* for short.

In the present chapter we will start to use substitution in Section 5.4, in a limited context where all terms are substitutable for all variable occurrences. At that point we will give a formal definition of (5.14) for this limited context. But the notion of 'substitutable for' will become important in Chapter 7, when we begin to use formal languages that have quantifier symbols.

Exercises

5.2.1. State the arities of the following predicates and terms.

(a) $x < y$.

(b) $x < y < z$.

(c) x is positive.

(d) $\lim_{x \to 0} \frac{h \sin x}{x}$.

5.2.2.   Suppose R is defined on the set $\{1, 2, 3, 4\}$ by

$$R(x, y, z) \;\Leftrightarrow\; x \text{ is a number strictly between } y \text{ and } z.$$

List all the 3-tuples in R.

5.2.3.   Consider how we define relations R on $\mathbb{R}$. Neither of the following definitions will work. Explain what goes wrong in each case.

(a) $R(x, y, y) \;\Leftrightarrow\; y = x + 2$.

(b) $R(x, y, z) \;\Leftrightarrow\; x + y = z + w$.

[In each case you should give a triple (a, b, c) of real numbers where the definition fails to tell us whether or not the triple is in the relation R.]

5.2.4.   How many different n-ary relations are there on a set of k elements?

5.2.5.   For each of the following expressions E, say whether the term $x^2 - yz$ is substitutable for the variable y in E. If it is, say what $E[x^2 - yz/y]$ is.

(a) y.

(b) z.

(c) $\int_y^z \sin(w) \, dw$.

(d) $\int_w^z \sin(y) \, dy$.

(e) $\int_1^2 (x + y) \, dx$.

5.2.6.   For each of the following expressions E, say whether the term $y + \cos(z)$ is substitutable for the variable x in E. If it is, say what $E[y + \cos(z)/x]$ is.

(a) x.

(b) $x = y$.

(c)   For all integers x, $5 \neq x^2$.

(d)   For all integers y, $5 \neq (x + y)^2$.

(e)   The set of reals z such that $|z - x| < 1$.

5.3   The language of first-order logic

We are going to introduce a language called LR (Language of Relations), that is much closer to normal mathematical language than LP was. For example, LR will have function and relation symbols. As with LP, the choice of signature is left to the user. We assume you have the good sense not to put parentheses or commas into your signature, since these symbols are needed for punctuation.

Definition 5.3.1 A *first-order signature* (shortened to *signature* when there is no danger of confusion with propositional signatures) is a 4-tuple (Co, Fu, Re, r) where:

(1) Co is a set (possibly empty) of symbols called the *constant symbols*;

(2) Fu is a set (possibly empty) of symbols called the *function symbols*;

(3) Re is a set (possibly empty) of symbols called the *relation symbols*;

(4) Co, Fu, Re are pairwise disjoint;

(5) r is a function taking each symbol s in $Fu \cup Re$ to a positive integer $r(s)$ called the *rank* (or *arity*) of s. We say a symbol is *n-ary* if it has arity n; *binary* means 2-ary.

Example 5.3.2

(a) The following signature will play an important role later and will serve as a concrete example for the present. The *signature of arithmetic*, σ_{arith}, has the following symbols:

    (i)   a constant symbol $\bar{0}$;

    (ii) a function symbol $\bar{S}$ of arity 1;

    (iii) two binary function symbols $\bar{+}$ and $\bar{\cdot}$.

We will usually present signatures in this informal style. To match Definition 5.3.1 we would put:

    (1) $Co_{\text{arith}} = \{\bar{0}\}$.

    (2) $Fu_{\text{arith}} = \{\bar{S}, \bar{+}, \bar{\cdot}\}$.

    (3) $Re_{\text{arith}} = \emptyset$.

    (4) $r_{\text{arith}}(\bar{S}) = 1$, $r_{\text{arith}}(\bar{+}) = r_{\text{arith}}(\bar{\cdot}) = 2$.

Then $\sigma_{\text{arith}} = (Co_{\text{arith}}, Fu_{\text{arith}}, Re_{\text{arith}}, r_{\text{arith}})$. Later we will use this signature to talk about the natural numbers. Then $\bar{0}$ will name the number 0, and the function symbols $\bar{+}$, $\bar{\cdot}$ will stand for plus and times. The symbol $\bar{S}$ will stand for the successor function $(n \mapsto n + 1)$.

(b) The signature of groups, σ_{group}, has a constant symbol and two function symbols, namely,

    (i)   a constant symbol 'e';

    (ii) a binary function symbol '$\cdot$';

    (iii) a 1-ary function symbol '$^{-1}$'.

(c) Neither of the previous signatures has any relation symbol; here is one that does. The signature of linear orders, σ_{lo}, has no constant symbols, no function symbols, and one binary relation symbol $<$.

Parsing

Suppose F is a 1-ary function symbol. Then the expression

(5.15) $$\exists x(F(F(x)) = x)$$

is a formula of LR, as will soon become clear. (It expresses that for some element x, $F(F(x))$ is equal to x.) If we parse it, we get the following parsing tree:

(5.16)

Just as in Chapter 3, we can reconstruct the formula by labelling the nodes of the parsing tree, starting at the leaves and working upwards.

The three nodes in the left-hand branch of the tree are the parsing tree of the term $F(F(x))$. It will make life simpler if we split the syntax of LR into two levels. The lower level consists of the *terms* of LR; the upper level consists of the *formulas*. Most formulas have terms inside them (the formula $\perp$ is one of the exceptions). But in first-order logic, terms never have formulas inside them. So we can describe the terms first, and then treat them as ingredients when we go on to build formulas. Thus we split (5.16) into two parsing trees:

(5.17)

The first is a parsing tree for a term and the second is a parsing tree for a formula.

Definition 5.3.3 (a) The *variables* of LR are the infinitely many symbols

$$x_0, x_1, x_2, \ldots$$

which we assume are not in σ.

(b) Let σ be a signature. A *parsing tree* for terms of $\mathrm{LR}(\sigma)$ is a right-labelled tree where

- if a node has arity 0 then its label is either a constant symbol of σ or a variable;
- if a node has arity $n > 0$ then its label is a function symbol of σ with arity n.

In the heat of battle we will often use x, y, z, etc. as variables of LR, because this is common mathematical notation. But strictly the variables of LR are just the variables in (a) above.

We can read a term from its parsing tree by the following compositional definition:

(5.18)

where α is a constant symbol or variable, and F is a function symbol of arity n.

For example, on the parsing tree

(5.19)

we build up the following labelling:

(5.20)

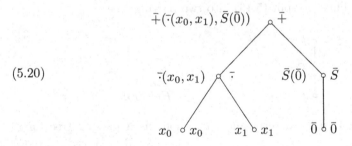

The left label on the root node is

(5.21)
$$\bar{+}(\bar{\cdot}(x_0, x_1), \bar{S}(\bar{0})).$$

So this is the term associated to the parsing tree (5.19). In normal mathematical notation we would usually write (5.21) as $x_0 x_1 + S(0)$; for purposes of logic it is helpful to think of $x_0 x_1 + S(0)$ as a shorthand for (5.21).

Definition 5.3.4 Let σ be a signature. Then a *term* of $\mathrm{LR}(\sigma)$ is the associated term of a parsing tree for terms of $\mathrm{LR}(\sigma)$. A *term* of LR is a term of $\mathrm{LR}(\sigma)$ for some signature σ.

For example, the following are terms of $\mathrm{LR}(\sigma_{\mathrm{arith}})$:

(5.22) $\qquad\qquad \bar{0} \qquad \bar{S}(\bar{S}(x_5)) \qquad \bar{+}(\bar{S}(\bar{0}), x_2) \qquad \bar{S}(\bar{\cdot}(x_4, \bar{0}))$

In normal mathematical shorthand these would appear as

(5.23) $\qquad\qquad\qquad 0 \qquad S(S(x_5)) \qquad S(0) + x_2 \qquad S(x_4 \cdot 0)$

Likewise when we do algebra in LR we should strictly write

$$(2 + 3x_0)^4$$

as

$$^4(+(2, \cdot(3, x_0)))$$

But in practice nobody can understand expressions like that, so we will generally stick to the usual notation, though we always regard it as a shorthand.

Formulas are more complicated than terms; there are more ways of building them up. To save repetition we will introduce formulas with quantifiers here, although we will not need them until the next chapter.

Definition 5.3.5 Let v be a variable. An expression $\forall v$ (read 'for all v') is called a *universal quantifier*. An expression $\exists v$ (read 'there is v') is called an *existential quantifier*. *Quantifiers* are universal quantifiers and existential quantifiers. (WARNING: Some logicians also refer to the quantifier symbols $\forall$ and $\exists$ as 'quantifiers'. We will avoid doing this.)

Definition 5.3.6 Let σ be a signature. A *parsing tree* for formulas of $\mathrm{LR}(\sigma)$ is a right-labelled tree where

- every leaf is labelled with either $\bot$ or a term of $\mathrm{LR}(\sigma)$;
- every node that is not a leaf is labelled with one of the symbols $=$, $\neg, \wedge, \vee, \rightarrow, \leftrightarrow$ or a relation symbol of σ, or a quantifier;
- every node labelled with a quantifier has arity 1, and its daughter node is not labelled with a term;
- every node labelled with $\neg$ has arity 1 and its daughter node is not labelled with a term;
- every node labelled with one of $\wedge, \vee, \rightarrow, \leftrightarrow$ has arity 2 and neither of its daughters is labelled with a term;

- if a node is labelled with =, its arity is 2 and its daughter nodes are labelled with terms;
- if a node is labelled with a relation symbol R, its arity is the arity of R and its daughter nodes are labelled with terms.

The compositional definition for building the associated formula of a parsing tree for formulas is as follows. It includes the clauses of (3.22) for the nodes labelled by truth function symbols, together with four new clauses for the leaves and the nodes labelled by a quantifier, '=' or a relation symbol:

$$Qv\phi \quad Qv$$

$$t \circ t \qquad\qquad \phi$$

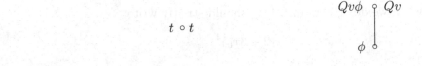

$$(5.24) \qquad (t_1 = t_2) \quad = \qquad\qquad R(t_1,\dots,t_n) \quad R$$

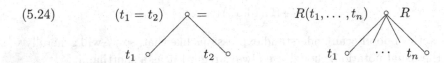

$$t_1 \qquad t_2 \qquad\qquad t_1 \quad \cdots \quad t_n$$

where t is a term, Qv is a quantifier and R is a relation
symbol of σ with arity n.

For example, here is an example of a parsing tree for a formula of $\mathrm{LR}(\sigma_{\mathrm{arith}})$:

$$\forall x$$

$$(5.25) \qquad\qquad \exists y$$

$$=$$

$$\bar{S}(x) \qquad \bar{\mp}(y,\bar{0})$$

Applying (5.24) to (5.25) yields the following left labelling (where the right-hand labels are omitted to save clutter):

$$\forall x \exists y (\bar{S}(x) = \bar{\mp}(y,\bar{0}))$$

$$\exists y (\bar{S}(x) = \bar{\mp}(y,\bar{0}))$$

$$(5.26) \qquad\qquad (\bar{S}(x) = \bar{\mp}(y,\bar{0}))$$

$$\bar{S}(x) \quad \bar{\mp}(y,\bar{0})$$

So the associated formula of (5.25), the label of the root, is

(5.27) $$\forall x \exists y (\bar{S}(x) = \bar{+}(y, \bar{0}))$$

In normal mathematical practice we would write (5.27) as

(5.28) $$\forall x \exists y \ (S(x) = y + 0)$$

As with terms, we regard (5.28) as shorthand for (5.27).

We note another very common shorthand: $s \neq t$ for the formula $\neg(s = t)$.

Definition 5.3.7 Let σ be a signature. Then a *formula* of $\mathrm{LR}(\sigma)$ is the associated formula of a parsing tree for formulas of $\mathrm{LR}(\sigma)$. A *formula* of LR is a formula of $\mathrm{LR}(\sigma)$ for some signature σ.

If the parsing trees were not of interest to us in their own right, we could boil down the definitions of 'term' and 'formula' to the following inductive definitions in the style of (3.5). First the terms:

(a) Every constant symbol of σ is a term of $\mathrm{LR}(\sigma)$.

(b) Every variable is a term of $\mathrm{LR}(\sigma)$.

(c) For every function symbol F of σ, if F has arity n and $t_1, \ldots, t_n$ are terms of $\mathrm{LR}(\sigma)$ then the expression $F(t_1, \ldots, t_n)$ is a term of $\mathrm{LR}(\sigma)$.

(d) Nothing is a term of $\mathrm{LR}(\sigma)$ except as implied by (a), (b), (c).

Then the formulas:

(a) If s, t are terms of $\mathrm{LR}(\sigma)$ then the expression $(s = t)$ is a formula of $\mathrm{LR}(\sigma)$.

(b) If R is a relation symbol of arity n in σ, and $t_1, \ldots, t_n$ are terms of $\mathrm{LR}(\sigma)$, then the expression $R(t_1, \ldots, t_n)$ is a formula of $\mathrm{LR}(\sigma)$.

(c) $\bot$ is a formula of $\mathrm{LR}(\sigma)$.

(d) If ϕ and ψ are formulas of $\mathrm{LR}(\sigma)$, then so are the following expressions:

$$(\neg\phi) \qquad (\phi \wedge \psi) \qquad (\phi \vee \psi) \qquad (\phi \rightarrow \psi) \qquad (\phi \leftrightarrow \psi)$$

(e) If ϕ is a formula of $\mathrm{LR}(\sigma)$ and v is a variable, then the expressions $\forall v \phi$ and $\exists v \phi$ are formulas of $\mathrm{LR}(\sigma)$.

(f) Nothing is a formula of $\mathrm{LR}(\sigma)$ except as implied by (a)–(e).

Unique parsing for LR

If someone gives us a term or a formula of $\mathrm{LR}(\sigma)$, then we can reconstruct its parsing tree by starting at the top and working downwards. This is intuitively clear, but as in Section 3.3 we want to prove it by showing a calculation that makes the construction automatic.

The first step is to find the *head*, that is, the symbol or quantifier that comes in at the root node. For terms this is easy: the head is the leftmost symbol in the term. If this symbol is a function symbol F, then the signature tells us its arity n, so we know that the term has the form $F(t_1, \ldots, t_n)$. But this does not tell us how to chop up the string $t_1, \ldots, t_n$ to find the terms t_1, t_2, etc. It is enough if we can locate the $n - 1$ commas that came with F, but this is not trivial since the terms t_1, etc. might contain commas too. Fortunately, the reasoning differs only in details from what we did already with LP, so we make it an exercise (Exercise 5.3.6). The corresponding reasoning for formulas is more complicated but not different in principle. The outcome is the following theorem, which we state without proof. It guarantees that each term or formula of $LR(\sigma)$ has a unique parsing tree.

Theorem 5.3.8 (Unique Parsing Theorem for LR) *Let σ be a signature. Then no term of $LR(\sigma)$ is also a formula of $LR(\sigma)$. If t is a term of $LR(\sigma)$, then exactly one of the following holds:*

(a) t is a variable.

(b) t is a constant symbol of $LR(\sigma)$.

(c) t is $F(t_1, \ldots, t_n)$ where n is a positive integer, F is a function symbol of σ with arity n, and $t_1, \ldots, t_n$ are terms of $LR(\sigma)$. Moreover, if t is also $G(s_1, \ldots, s_m)$ where G is an m-ary function symbol of $LR(\sigma)$ and $s_1, \ldots, s_m$ are terms of $LR(\sigma)$, then $n = m$, $F = G$ and each t_i is equal to s_i.

If ϕ is a formula of $LR(\sigma)$, then exactly one of the following is true:

(a) ϕ is $R(t_1, \ldots, t_n)$ where n is a positive integer, R is a relation symbol of σ with arity n, and $t_1, \ldots, t_n$ are terms of $LR(\sigma)$. Moreover, if ϕ is also $P(s_1, \ldots, s_m)$ where P is an m-ary relation symbol of $LR(\sigma)$ and $s_1, \ldots, s_m$ are terms of $LR(\sigma)$, then $n = m$, $R = P$ and each t_i is equal to s_i.

(b) ϕ is $(s = t)$ for some terms s and t of $LR(\sigma)$. Moreover, if ϕ is also $(s' = t')$ where s' and t' are terms of $LR(\sigma)$, then s is s' and t is t'.

(c) ϕ is $\bot$.

(d) ϕ has exactly one of the forms $(\phi_1 \wedge \phi_2)$, $(\phi_1 \vee \phi_2)$, $(\phi_1 \rightarrow \phi_2)$, $(\phi_1 \leftrightarrow \phi_2)$, where ϕ_1 and ϕ_2 are uniquely determined formulas.

(e) ϕ is $(\neg \psi)$ for some uniquely determined formula ψ.

(f) ϕ has exactly one of the forms $\forall x \psi$ and $\exists x \psi$, where x is a uniquely determined variable and ψ is a uniquely determined formula.

In the light of this theorem, we can define properties of terms and formulas by using properties of their parsing trees. For clauses (b) and (c) below, recall Definition 3.2.9.

Definition 5.3.9

(a) The *complexity* of a term or formula is the height of its parsing tree. A term or formula with complexity 0 is said to be *atomic*; all other terms and formulas are said to be *complex*.

(b) Let t be a term of LR. Then the *subterms* of t are the traces in t of the left labels on the nodes of the parsing tree of t. The *proper subterms* of t are all the subterms except t itself. The *immediate subterms* of t are the traces coming from the daughters of the root of the parsing tree.

(c) Let ϕ a formula of LR. Then the *subformulas* of ϕ are the traces in ϕ of the left labels on the nodes of the parsing tree of ϕ. The *proper subformulas* of ϕ are all the subformulas except ϕ itself. The *immediate subformulas* of ϕ are the traces coming from the daughters of the root of the parsing tree.

(d) Let ϕ be a formula of LR. We say that ϕ is *quantifier-free* if $\forall$ and $\exists$ never occur in ϕ (or equivalently, if no nodes of the parsing tree of ϕ are right labelled with quantifiers). We say *qf formula* for *quantifier-free formula*. We say *qf* LR to mean the terms and qf formulas of LR.

By this definition the atomic terms are exactly those that have no proper subterms; in other words, they are the variables and the constant symbols. Likewise the atomic formulas are exactly those with no proper subformulas; in other words, they are $\perp$ and formulas of the forms $(s = t)$ or $R(t_1, \ldots, t_n)$. Every atomic formula is quantifier-free. But there are qf formulas that are not atomic; the smallest is the formula $(\neg\perp)$.

For example, the expression

$$\bar{+}(\bar{S}(\bar{S}(\bar{0})), \bar{+}(\bar{0}, x_1))$$

is a term of $\mathrm{LR}(\sigma_{\mathrm{arith}})$, and it has the parsing tree

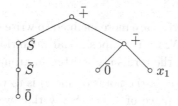

It is a complex term with complexity 3. It has seven subterms, as follows:

$$\bar{+}(\bar{S}(\bar{S}(\bar{0})), \bar{+}(\bar{0}, x_1))$$

———————————— (immediate subterm)

———

———————— (immediate subterm)

–

Exercises

5.3.1. Suppose σ is a signature. What are the symbols in the lexicon of LR(σ)? (Cf. the introduction to Chapter 3, and Definition 3.1.1(a) for propositional logic.)

5.3.2. Determine which of the following expressions are terms of LR(σ_{arith}). For those which are, give parsing trees. For those which are not, say why they are not. (Shorthand terms are not allowed here.)

   (a) $\bar{S}\bar{S}\bar{0}$.

   (b) $\bar{\cdot}(\bar{S}(\bar{0}), \bar{x})$.

   (c) $\bar{+}(\bar{\cdot}(y, \bar{S}(x)), \bar{S}(x), \bar{0})$.

   (d) $\bar{S}(\bar{\cdot}(\bar{+}(x, \bar{S}(\bar{S}(\bar{0}))), \bar{+}(y, x)))$.

5.3.3. Determine which of the following expressions are formulas of LR(σ_{arith}). For those which are, give parsing trees. For those which are not, say why they are not. (Shorthand formulas are not allowed here.)

   (a) $(\bar{+}(\bar{S}(\bar{S}(\bar{0})), \bar{S}(\bar{S}(\bar{0}))) = \bar{0})$.

   (b) $\forall x_1(\exists x_2(x_1 = x))$.

   (c) $\exists x_1 \forall x_2 \forall x_3((x_1 = \bar{\cdot}(x_2, x_3)) \rightarrow ((x_2 = \bar{0}) \vee (x_3 = \bar{0})))$.

   (d) $(\bar{+}(\bar{S}(\bar{0}), \bar{S}(\bar{0})) = \bar{S}(\bar{S}(\bar{0}))))$.

5.3.4. Identify all the subterms of the following term of LR(σ_{group}). Say which are immediate subterms, and which are atomic.

$$\cdot (e, \cdot ({}^{-1}(y), {}^{-1}({}^{-1}(x))))$$

5.3.5. (a) Group theorists are more inclined to write $\cdot(x, y)$ as $(x \cdot y)$ and ${}^{-1}(x)$ as $(x)^{-1}$. Write a compositional definition that works like (5.18) except that the terms are written in group theorists' notation.

   (b) One variation of the notation in (a) is to write xy in place of $\cdot(x, y)$ and x^{-1} in place of ${}^{-1}(x)$. Show that with this notation, unique parsing fails. When the notation is used for groups, does the failure of unique parsing matter?

5.3.6. (a) Let the signature σ include function symbols F, G, H of arities 3, 3 and 2, respectively, and a relation symbol R of arity 4. Then

$$R(G(H(x, x), y, F(x, y, z)), H(G(u, v, u), v), w, F(y, H(u, H(v, v)), w))$$

is an atomic formula of LR(σ). Calculate the depths of all the commas in this formula (where depth is defined as in Definition 3.3.1).

(b) Formulate and prove a theorem about how to find the three commas associated with the relation symbol R.

5.4 Proof rules for equality

Atomic formulas of the form $(s = t)$ are called *equations*, and the symbol '$=$' is known as *equality* or *identity*. Some logicians regard it as a binary relation symbol that needs to be mentioned in the signature, but for us the formulas of LR(σ) always include equations, regardless of what σ is.

The natural deduction rules for qf formulas are exactly the same as for LP, except that we also have introduction and elimination rules ($=$I), ($=$E) for equality. This may be a good place to remark that the symbols s and t in the derivations of this section are metavariables ranging over terms. So, for example, (5.29) is not a derivation but a pattern for infinitely many derivations, depending on what terms of LR we put for s and t.

NATURAL DEDUCTION RULE ($=$I) If t is a term then

$$(5.29) \qquad \frac{}{(t = t)}\ (=I)$$

is a derivation of the formula $(t = t)$. It has no assumptions.

The line above $(t = t)$ indicates that this is a derivation of $(t = t)$ from no assumptions, not a derivation of $(t = t)$ from $(t = t)$ using the Axiom Rule.

Example 5.4.1 At first sight it seems unrealistic to think that we would ever start a proof by asserting that $0 = 0$. But rules for manipulating formulas have to be taken as a package; this rule will interact with the other rule for equality. In any case, one can construct realistic arguments which use the fact that $0 = 0$.

For example, suppose the function f is defined by

$$f(x) = \frac{x^2 + 2x}{x}$$

Strictly this function is undefined at $x = 0$ since you cannot divide by 0. But there is an obvious way of extending it to be defined on all reals:

$$g(x) = \begin{cases} f(x) & \text{if } x \neq 0, \\ 2 & \text{if } x = 0. \end{cases}$$

Now we calculate $g(0)$ from this definition:

$0 = 0$, so the second line of the definition applies, and $g(0) = 2$.

Note how we used the fact that $0 = 0$ in order to apply a case of the definition.

The second rule of equality uses substitutions (as in Definition 5.2.10).

NATURAL DEDUCTION RULE (=E) If ϕ is a formula, s and t are terms substitutable for x in ϕ, and

$$
\begin{array}{cc}
D & D' \\
(s = t)\ ' & \phi[s/x]
\end{array}
$$

are derivations, then so is

$$
(5.30) \qquad \frac{\begin{array}{cc} D & D' \\ (s = t) & \phi[s/x] \end{array}}{\phi[t/x]}\ (=E)
$$

Its undischarged assumptions are those of D together with those of D'. (The reference to substitutability will not make any difference until Chapter 7, because if ϕ is quantifier-free then any term is substitutable for x in ϕ.)

The rule (=E) is sometimes called *Leibniz's Law*. The idea behind it is that if s and t are the identically same object, and something is true about s, then the same thing is true about t. This rule has an important new feature. The formula $\phi[s/x]$ does not by itself tell you what ϕ, s and x are, because '[/]' is our notation for substitutions, not part of the formula. The same applies with $\phi[t/x]$. So to check that you do have an example of (5.30), you have to find appropriate ϕ, s, t and x so that $\phi[s/x]$ and $\phi[t/x]$ work out as the required formulas. (Exercise 5.4.3 will give you some practice.)

Example 5.4.2 We prove the sequent $\{(s = t)\} \vdash (t = s)$ by the following derivation:

$$
(5.31) \qquad \frac{(s = t) \qquad \dfrac{}{(s = s)}\ (=I)}{(t = s)}\ (=E)
$$

Above the central line in (5.31), the derivation on the left (which is D in the notation of (5.30)) consists of the formula $(s = t)$; it is both conclusion and undischarged assumption. The derivation on the right (the D' of (5.30)) is an application of (=I). The bottom line of (5.31) follows by an application of (=E), but we have to analyse a little in order to see this. We have to find a qf formula ϕ such that

$$
(5.32) \qquad \begin{array}{l} \phi[s/x] \text{ is the formula } (s = s) \\ \phi[t/x] \text{ is the formula } (t = s) \end{array}
$$

A small amount of pencil-and-paper work shows that if ϕ is the formula $(x = s)$, then (5.32) holds, completing the proof.

Suppose now that D is a derivation with conclusion $(s = t)$. Then the following is a derivation of $(t = s)$, with exactly the same undischarged assumptions as D:

$$(5.33) \qquad \frac{\begin{array}{cc} \begin{array}{c} D \\ (s = t) \end{array} & \dfrac{}{(s = s)}\ (\text{=I}) \end{array}}{(t = s)}\ (\text{=E})$$

It follows that we could without any qualms introduce another natural deduction rule (=symmetric) as follows:

If s, t are terms and

$$\begin{array}{c} D \\ (s = t) \end{array}$$

is a derivation then so is

$$(5.34) \qquad \frac{\begin{array}{c} D \\ (s = t) \end{array}}{(t = s)}\ (\text{=symmetric})$$

This rule gives us nothing new, because every time we use (5.34) in a derivation, we can rewrite it as (5.33) without affecting either the conclusion or the undischarged assumptions. However, (5.34) has one line less than (5.33), and this is a bonus.

Here are two more rules that we can justify in similar ways to (=symmetric). The justifications are left to Exercise 5.4.5. (And what are the undischarged assumptions of the constructed derivations?)

(1) **The rule (=transitive)**: If s, t, u are terms and

$$\begin{array}{c} D \\ (s = t) \end{array} \quad \text{and} \quad \begin{array}{c} D' \\ (t = u) \end{array}$$

are derivations then so is

$$(5.35) \qquad \frac{\begin{array}{cc} \begin{array}{c} D \\ (s = t) \end{array} & \begin{array}{c} D' \\ (t = u) \end{array} \end{array}}{(s = u)}\ (\text{=transitive})$$

(2) **The rule (=term)**: If r, s, t are terms and

$$\begin{array}{c} D \\ (s = t) \end{array}$$

is a derivation then so is

(5.36)
$$\frac{\begin{array}{c} D \\ (s = t) \end{array}}{r[s/x] = r[t/x]} \quad (=\text{term})$$

Example 5.4.3 Consider a group G with identity element e, and write xy for the product of the elements x and y. Then the equations

$$yx = e, \qquad xz = e$$

imply $y = z$. The usual proof is brief and to the point:

$$y = ye = y(xz) = (yx)z = ez = z.$$

The natural deduction proof is longer but it does cover the same ground. The derivation below uses a more relaxed notation than $\text{LR}(\sigma_{\text{group}})$, but the applications of the natural deduction rules (including the derived rules (=symmetric), (=transitive) and (=term)) are strictly accurate. Besides the assumptions $yx = e$ and $xz = e$, the derivation uses assumptions that are guaranteed by the definition of 'group', for example, $ye = y$ and $y(xz) = (yx)z$.

(5.37)
$$\frac{\dfrac{ye = y}{y = ye} \quad \dfrac{\dfrac{xz = e}{e = xz}}{ye = y(xz)} \quad \dfrac{\dfrac{y(xz) = (yx)z \quad \dfrac{\dfrac{yx = e}{(yx)z = ez} \quad ez = z}{(yx)z = z}}{y(xz) = z}}{ye = z}}{y = z}$$

Exercise 5.4.6 invites you to say which rule is used at each step in (5.37).

Formalising natural deduction for qf LR

To formalise the natural deduction calculus, our first step must be to give the promised formalisation of substitution (Definition 5.2.10) for the case of terms and qf formulas. In this case, there is nothing to bind variables, so every term is substitutable for every variable. This allows us to copy Definition 3.7.3 by recursion on complexity. (Conscientious logicians prove, by induction on complexity, that in each case below $E[t_1/y_1, \ldots, t_k/y_k]$ really is a term or a formula as required; but we will take this result as obvious.)

Definition 5.4.4 Let E be a term or qf formula, $y_1, \ldots, y_n$ distinct variables and $t_1, \ldots, t_n$ terms.

(a) If E is a variable or a constant symbol, then

$$E[t_1/y_1, \ldots, t_k/y_k] \text{ is } \begin{cases} t_i & \text{if } E \text{ is } y_i \ (1 \leqslant i \leqslant k), \\ E & \text{otherwise.} \end{cases}$$

(b) If E is $F(s_1, \ldots, s_n)$ where F is a function symbol of arity n, then $E[t_1/y_1, \ldots, t_k/y_k]$ is

$$F\left(s_1[t_1/y_1, \ldots, t_k/y_k], \ldots, s_n[t_1/y_1, \ldots, t_k/y_k]\right).$$

The same applies with a relation symbol R in place of the function symbol F. (The spacing above is just for readability.)

(c) If E is $(s_1 = s_2)$ then $E[t_1/y_1, \ldots, t_k/y_k]$ is $(s_1[t_1/y_1, \ldots, t_k/y_k] = s_2[t_1/y_1, \ldots, t_k/y_k])$.

(d) If E is $(\neg\phi)$ where ϕ is a formula, then $E[t_1/y_1, \ldots, t_k/y_k]$ is $(\neg\phi[t_1/y_1, \ldots, t_k/y_k])$. Similar clauses apply when E is $(\phi \wedge \psi)$, $(\phi \vee \psi)$, $(\phi \rightarrow \psi)$ or $(\phi \leftrightarrow \psi)$.

Next, we adapt to qf LR our definition of σ-derivations (Definition 3.4.1).

Definition 5.4.5 Let σ be a first-order signature. The definition of σ-*derivation for qf LR* is the same as the definition of σ-derivation in Definition 3.4.1 with the following changes. We have qf $\mathrm{LR}(\sigma)$ in place of $\mathrm{LP}(\sigma)$ throughout. Clause (c) becomes:

(c) Every node of arity 0 carries as right-hand label either (A) or (=I).

Under (e) we have the new clause:

(v) ν has right-hand label (=E), and for some formula ϕ, variable x and terms s, t, the left labels on the daughters of ν are (from left to right) $(s = t)$ and $\phi[s/x]$, and the left label on ν is $\phi[t/x]$.

The definition of *conclusion* is as before, but the *undischarged assumptions* are now the formulas that appear without dandahs as left labels on leaves labelled (A).

Theorem 5.4.6 *Let σ be a finite signature. There is an algorithm that, given any diagram, will determine in a finite amount of time whether or not the diagram is a σ-derivation for qf LR.*

The proof follows that of Theorem 3.4.2, but using Definition 5.4.5.

Definition 5.4.7

(a) The definitions of *qf σ-sequent*, *conclusion* of a sequent, *assumptions* of a sequent and *correctness* of a sequent, and of a derivation *proving* a sequent, are all as in Definition 3.4.4 with the obvious changes from LP to qf LR.

(b) A *derived rule* of natural deduction is a possible natural deduction rule that is not among the rules (Axiom Rule), ($\wedge$I), ($\wedge$E), ($\vee$I), ($\vee$E), ($\rightarrow$I), ($\rightarrow$E), ($\leftrightarrow$I), ($\leftrightarrow$E), ($\neg$I), ($\neg$E), (RAA), (=I), (=E) (or the rules ($\forall$I), ($\forall$E), ($\exists$I), ($\exists$E) to be introduced in Chapter 7), but is valid in the sense that if it is added to those rules, no new sequents become provable. (So (=symmetric) is a derived rule.)

(c) A set Γ of qf sentences of LR(σ) is *syntactically consistent* if $\Gamma \nvdash_\sigma \bot$, that is, if there is no derivation of $\bot$ whose undischarged assumptions are in Γ.

As before, when the context allows, we leave out σ and write $\vdash_\sigma$ as $\vdash$. Exercise 5.4.7 explains why the choice of signature σ is irrelevant so long as qf LR(σ) includes the assumptions and conclusion of the sequent.

It remains to state the two sequent rules that describe what sequents are proved by the natural deduction rules (=I) and (=E):

SEQUENT RULE (=I) For every term t, the sequent ($\vdash (t = t)$) is correct.

SEQUENT RULE (=E) For every pair of terms s and t, every variable x, every formula ϕ and all sets of formulas Γ, Δ, if s and t are both substitutable for x in ϕ, and the sequents ($\Gamma \vdash \phi[s/x]$) and ($\Delta \vdash (s = t)$) are both correct, then ($\Gamma \cup \Delta \vdash \phi[t/x]$) is a correct sequent.

Exercises

5.4.1. Evaluate the following:

(a) $\mp(x_1, x_2)[x_2/x_1, \bar{0}/x_2]$.

(b) $\mp(x_1, x_2)[x_2/x_2, \bar{0}/x_1]$.

(c) $\mp(x_1, x_2)[\bar{0}/x_2, \bar{S}(\bar{0})/x_3]$.

(d) $(\mp(x_1, x_2) = x_0)[x_2/x_1][\bar{0}/x_1, \bar{S}(\bar{0})/x_2]$.

5.4.2. In LR(σ_{arith}):

(a) Find a term t such that $\mp(\bar{S}(x), x)[t/x]$ is $\mp(\bar{S}(\bar{\cdot}(x, \bar{S}(x))), \bar{\cdot}(x, \bar{S}(x)))$.

(b) Find a term t such that $(\bar{S}(x) = \bar{S}(\bar{S}(y)))[t/y]$ is the formula $(\bar{S}(x) = \bar{S}(\bar{S}(\bar{\cdot}(y, y))))$.

(c) Find a term t in which x occurs, such that $t[\mp(y, y)/x]$ is $\mp(\mp(y, y), y)$.

(d) Find a formula ϕ in which $\bar{0}$ occurs exactly once, such that $\phi[\bar{0}/z]$ is $(\bar{0} = {}^{\textstyle\cdot}(\bar{0}, \bar{0}))$. [There is more than one possible answer.]

5.4.3. In $\mathrm{LR}(\sigma_{\mathrm{arith}})$ but using relaxed notation:

(a) Find a formula ϕ and terms s, t so that

$$\phi[s/x] \text{ is } SSS0 = z \text{ and } \phi[t/x] \text{ is } SSSS0 = z.$$

(b) Find a formula ϕ and terms s, t so that

$$\phi[s/x] \text{ is } y = x + z \text{ and } \phi[t/x] \text{ is } y = z + z.$$

(c) Find a formula ϕ and terms s, t so that

$$\phi[s/x, t/y] \text{ is } (0 + SS0) + SSSS0 = SS(0 + SS0) \text{ and}$$
$$\phi[t/x, s/y] \text{ is } SSS0 + S(0 + SS0) = SSSSS0.$$

5.4.4. The following derivation is part of a proof that $2 + 2 = 4$, but consider it simply as a derivation using formulas of $\mathrm{LR}(\sigma_{\mathrm{arith}})$.

$$\cfrac{\cfrac{SS0 + 0 = SS0 \qquad SS0 + S0 = S(SS0 + 0)}{SS0 + S0 = SSS0}\ (=\mathrm{E}) \qquad SS0 + SS0 = S(SS0 + S0)}{SS0 + SS0 = SSSS0}\ (=\mathrm{E})$$

(i) Explain what substitutions are used at each of the applications of $(=\mathrm{E})$.

(ii) Show how to extend the derivation to one whose conclusion is $SS0 + SSS0 = SSSSS0$, using the assumption $SS0 + SSS0 = S(SS0 + SS0)$ in addition to those already used.

5.4.5. Give derivations to show that all sequents for qf LR of the following forms are correct, where r, s, t, u are any terms of LR.

(a) $\{(s = t), (t = u)\} \vdash (s = u)$.

(b) $\{(s = t)\} \vdash (r[s/x] = r[t/x])$.

Use your results to justify the derived rules $(=\text{transitive})$ and $(=\text{term})$.

5.4.6. Consider the derivation (5.37). Say which rule is used at each step. For the applications of $(=\text{term})$, say what substitutions are made.

5.4.7. Let σ be a signature, r a term of qf $\mathrm{LR}(\sigma)$, y a variable and ϕ a formula of qf $\mathrm{LR}(\sigma)$. Let D be a qf σ-derivation, and write D' for the diagram got from D by replacing each occurrence of the formula ϕ in D by $\phi[r/y]$. Show that D' is also a qf σ-derivation. [A full proof will check that D' satisfies every clause of Definition 5.4.5. But since the argument is nearly the same in most cases, you might take a representative sample. It should include (e)(v), the case of $(=\mathrm{E})$.]

5.4.8. Suppose ρ and σ are LR signatures with $\rho \subseteq \sigma$. Let Γ be a set of formulas of qf LR(ρ) and ψ a formula of qf LR(ρ). Show that the sequent $(\Gamma \vdash_\rho \psi)$ is correct if and only if the sequent $(\Gamma \vdash_\sigma \psi)$ is correct. [From left to right is essentially the same as Exercise 3.4.3. From right to left is like Exercise 3.4.4, but longer because now we have more kinds of symbols. Remove atomic formulas beginning with a relation symbol not in ρ by replacing them by $\bot$. Remove a function symbol F not in ρ by replacing each occurrence of a term $F(s_1, \ldots, s_n)$ (even inside another term) by the variable x_0; this case needs checking with care. The method that works for function symbols works also for constant symbols. In each of these cases, write t', ϕ' for the term or formula that results from making the replacement in a term t or a formula ϕ. You can give a recursive definition of t' or ϕ' just as in Definition 5.4.4.]

5.5 Interpreting signatures

George Peacock England, 1791–1858.
We can calculate with mathematical symbols, and then 'interpret' them afterwards.

The truth function symbols $\bot$, $\wedge$, etc. in LR mean the same as they meant in LP. Also we know what '=' means. But the symbols in a signature are just symbols; they do not mean anything. As we said before, the formulas of LR are not statements so much as patterns of statements.

This is not something peculiar to logic. When a group theorist proves

$$(x \cdot y)^{-1} = y^{-1} \cdot x^{-1} \text{ for all } x, y$$

the symbols '·' and '-1' are meaningless until we make them refer to a particular group. So we can say to the group theorist 'Give me an example', and the group theorist gives us a group. In other words, she tells us (1) what are the group elements, (2) what counts as multiplying them and (3) what the inverse function is.

Exactly the same holds with signatures in general. To interpret a signature σ, we provide a mathematical structure called a *σ-structure*. The same signature σ has many different σ-structures, just as group theory has many different groups.

Example 5.5.1 Recall the signature σ_{arith} of Example 5.3.2(a). One possible way of interpreting the symbols in σ_{arith} is to write the following chart:

(5.38)

domain	the set of all natural numbers $\{0, 1, 2, \ldots\}$;
$\bar{0}$	the number 0;
$\bar{S}(x)$	$x + 1$;
$\bar{+}(x, y)$	$x + y$;
$\bar{\cdot}(x, y)$	xy.

We analyse this chart, starting at the top row.

When we discuss a topic, some things are relevant to the topic and some are not. The collection of all relevant things is called the *domain*. The members of the domain are known as the *elements* of the structure; some logicians call them the *individuals*. So the first line of (5.38) fixes the subject-matter: σ_{arith} will be used to talk about natural numbers.

Thus the domain is a set of objects. The remaining lines of (5.38) attach features of this set to the various symbols in the signature. The symbol $\bar{S}$, for example, is a function symbol of arity 1, and (5.38) attaches to it the 1-ary function defined by the term $x + 1$. (See Section 5.2 for the use of terms to define functions.) Likewise the chart assigns 2-ary functions to $\bar{+}$ and $\bar{\cdot}$ by giving terms that define the functions. It tells us that $\bar{+}$ is interpreted as addition on the natural numbers, and $\bar{\cdot}$ as multiplication.

Definition 5.5.2 Given a signature σ, a *σ-structure* A consists of the following ingredients:

(a) A domain, which is a non-empty set. We refer to the elements of the domain as the *elements* of A.

(b) For each constant symbol c of σ, an element of the domain; we write this element as c_A.

(c) For each function symbol F of σ with arity n, an n-ary function on the domain; we write this function as F_A.

(d) For each relation symbol R of σ with arity n, an n-ary relation on the domain; we write this relation as R_A.

Example 5.5.1 (continued) The σ_{arith}-structure in Definition 5.5.2 is known as $\mathbb{N}$, the *natural number structure*. Then $\bar{+}_{\mathbb{N}}$ is the function $+$ on the natural numbers, and $\bar{\cdot}_N$ is multiplication of natural numbers. The element $\bar{0}_N$ is the number 0, and $\bar{S}_N$ is the *successor function* $(n \mapsto n + 1)$. (The symbol $\mathbb{N}$ is also used for the *set* of natural numbers, that is, the *domain* of the structure $\mathbb{N}$. In practice this ambiguity does not cause serious problems.)

Our next example is a different interpretation of the same signature σ_{arith}.

Example 5.5.3 Let m be a positive integer. Then $\mathbb{N}/m\mathbb{N}$, the structure of integers mod m, is defined to be the following σ_{arith}-structure.

$$(5.39)$$

domain	the set of natural numbers $\{0, 1, \dots, m - 1\}$;
$\bar{0}$	the number 0;
$\bar{S}(x)$	$\begin{cases} x + 1 & \text{if } x < m - 1; \\ 0 & \text{if } x = m - 1; \end{cases}$
$\bar{+}(x, y)$	the remainder when $x + y$ is divided by m;
$\bar{\cdot}(x, y)$	the remainder when xy is divided by m.

Consider the following formula of $\text{LR}(\sigma_{\text{arith}})$:

$$(5.40) \qquad (\bar{S}(\bar{0}) = \bar{S}(\bar{S}(\bar{S}(\bar{S}(\bar{0})))))$$

or in a natural shorthand

$$S0 = SSSS0.$$

Interpreted in $\mathbb{N}$, this says

$$0 + 1 = (((0 + 1) + 1) + 1) + 1$$

In other words, $1 = 4$. This is a false statement. But in $\mathbb{N}/3\mathbb{N}$ the calculation is different:

$$(5.41)$$

$S0$	means	$0 + 1,$	i.e.	1
$SS0$	means	$1 + 1,$	i.e.	2
$SSS0$	means	$2 + 1,$	i.e.	0
$SSSS0$	means	$0 + 1,$	i.e.	1

So in $\mathbb{N}/3\mathbb{N}$, the formula (5.40) makes the true statement that $1 = 1$.

Comparing $\mathbb{N}$ and $\mathbb{N}/m\mathbb{N}$ yields three morals:

(1) Two very different structures can have the same signature.

(2) The same formula can be true in one structure and false in another.

(3) Sometimes the same element is named by many different terms. For example in $\mathbb{N}/3\mathbb{N}$ the element 1 is named by

$$S0, \ SSSS0, \ SSSSSSS0, \ SSSSSSSSSS0, \ \dots$$

Bernard Bolzano Bohemia, 1781–1848.
The truth value of 'It's snowing' depends on the time and
place where it is used.

In calculating the value of $SSSS0$ in $\mathbb{N}$, we used some commonsense assumptions, for example, that if the function S is written four times then it is supposed to be applied four times. This is exactly the kind of assumption that logicians need to make explicit, because it will be needed in proofs of logical theorems. Section 5.6 will develop this point.

Neither of the two charts (5.38) and (5.39) has any relation symbols, because their signatures had no relation symbols. Here is another signature that does have one.

Definition 5.5.4 The signature σ_{digraph} has just one symbol, the binary relation symbol E. A σ_{digraph}-structure is known as a *directed graph*, or a *digraph* for short. Suppose G is a directed graph. The elements of the domain of G are called its *vertices*. The binary relation E_G is called the *edge relation* of G, and the ordered pairs in this relation are called the *edges* of G. The convention is to draw the vertices as small circles, then to draw an arrow from vertex μ to vertex ν if and only if the ordered pair (μ, ν) is in the edge relation of G.

For example, here is a digraph:

(5.42)

The following chart is the digraph (5.42) written as an $\text{LR}(\sigma_{\text{digraph}})$-structure.

$$(5.43) \quad \begin{array}{c|l} \text{domain} & \{1, 2, 3, 4\}; \\ E(x, y) & \{(1, 1), (1, 2), (2, 3), (3, 1)\}. \end{array}$$

Note how we name the edge relation by listing it in the style of (5.6). For some other digraphs it would be more sensible to name the edge relation by using a predicate.

Exercises

5.5.1. Draw the following digraphs:

(a)

$$(5.44) \quad \begin{array}{r|l} \text{domain} & \{1,2,3,4,5\}; \\ E(x,y) & y \text{ is either } x+1 \text{ or } x-4. \end{array}$$

(b)

$$(5.45) \quad \begin{array}{r|l} \text{domain} & \{1,2,3,4,5\}; \\ E(x,y) & \{(1,1),(1,4),(1,5),(2,2),(2,3),(3,2),(3,3), \\ & (4,1),(4,4),(4,5),(5,1),(5,4),(5,5)\}. \end{array}$$

5.5.2. Another way to describe a digraph is to introduce new constant symbols to stand for its vertices, and then write down a description in the resulting language $\mathrm{LR}(\sigma')$ (where σ' is σ_{digraph} together with the added symbols). For example, to describe a digraph with five vertices 1, 2, 3, 4, 5, we can introduce symbols $\bar{1}$, $\bar{2}$, $\bar{3}$, $\bar{4}$ and $\bar{5}$ to name these vertices, and then write the qf formula

$$\begin{aligned} (E(\bar{1},\bar{1}) \wedge \quad &\neg E(\bar{1},\bar{2}) \wedge \quad \neg E(\bar{1},\bar{3}) \wedge \quad E(\bar{1},\bar{4}) \wedge \quad \neg E(\bar{1},\bar{5}) \wedge \\ \neg E(\bar{2},\bar{1}) \wedge \quad &E(\bar{2},\bar{2}) \wedge \quad E(\bar{2},\bar{3}) \wedge \quad \neg E(\bar{2},\bar{4}) \wedge \quad E(\bar{2},\bar{5}) \wedge \\ \neg E(\bar{3},\bar{1}) \wedge \quad &E(\bar{3},\bar{2}) \wedge \quad E(\bar{3},\bar{3}) \wedge \quad \neg E(\bar{3},\bar{4}) \wedge \quad E(\bar{3},\bar{5}) \wedge \\ E(\bar{4},\bar{1}) \wedge \quad &\neg E(\bar{4},\bar{2}) \wedge \quad \neg E(\bar{4},\bar{3}) \wedge \quad E(\bar{4},\bar{4}) \wedge \quad \neg E(\bar{4},\bar{5}) \wedge \\ \neg E(\bar{5},\bar{1}) \wedge \quad &E(\bar{5},\bar{2}) \wedge \quad E(\bar{5},\bar{3}) \wedge \quad \neg E(\bar{5},\bar{4}) \wedge \quad E(\bar{5},\bar{5})) \end{aligned}$$

(We could also add $\bar{1} \neq \bar{2} \wedge \bar{1} \neq \bar{3} \wedge \dots$, but here let us take that as assumed.) Draw the digraph that has this description.

5.5.3. Here is the multiplication table of a cyclic group of order 3:

	e	a	b
e	e	a	b
a	a	b	e
b	b	e	a

We can rewrite this multiplication in the style of Exercise 5.5.2. For this we take the cyclic group as a σ_{group}-structure (see Example 5.3.2(b)), and we add to the signature two constant symbols $\bar{a}$ and $\bar{b}$ to serve as names for a and b (σ_{group} already gives us the name e for the identity).

Then we write the conjunction of all the true equations expressed by the multiplication table (such as $\bar{a} \cdot \bar{b} = e$). Write out this conjunction.

5.5.4. Consider the σ_{group}-structure

(5.46)

domain	the interval $(-1, 1)$ in the real numbers;
e	0;
$x \cdot y$	$\frac{x+y}{1+xy}$;
x^{-1}	$-x$.

(WARNING: The chart is giving new meanings to the symbols on the left. The symbols on the right have their usual meanings in the real numbers—otherwise how would we understand the chart?)

(a) Is this σ_{group}-structure a group? Give reasons.

(b) Using the new constant symbol $\bar{r}$ to name a real number r, which of the following formulas are true in the structure above, and which are false? Show your calculations.

(i) $\overline{1/3} \cdot \overline{1/3} = \overline{1/9}$.

(ii) $\overline{1/3} \cdot \overline{1/3} = \overline{2/3}$.

(iii) $\overline{1/2} \cdot \overline{1/2} = \overline{4/5}$.

(iv) $\overline{1/4}^{-1} \cdot \overline{-1/4} = e$.

(v) $\overline{1/2} \cdot \overline{-1/3} = \overline{1/5}$.

(The structure in this exercise describes addition of velocities in special relativity.)

5.5.6 Suppose π is a permutation of the set $\{1, 2, 3, 4, 5\}$, and G is the digraph of Exercise 5.5.1(b). We can apply π to G by putting $\pi(1)$ in place of 1 in G, $\pi(2)$ in place of 2 and so on.

(a) Find a permutation π that turns G into the digraph of Exercise 5.5.2.

(b) Without going through all the possible permutations, explain why there could not be a permutation that turns G into the digraph of Exercise 5.5.1(a).

5.5.7. This exercise generalises Exercise 5.5.6. Suppose σ is a signature and A, B are two σ_{digraph}-structures. An *isomorphism* from A to B is a bijection f from the domain of A to the domain of B such that for any two elements a, a' of A,

$$(a, a') \in E_A \iff (f(a), f(a')) \in E_B.$$

An isomorphism from A to A is called an *automorphism* of A.

(a) How many automorphisms does the digraph of Exercise 5.5.1(a) have?

(b) How many automorphisms does the digraph of Exercise 5.5.1(b) have?

(c) Draw a digraph with four vertices which has no automorphisms except the identity permutation.

5.6 Closed terms and sentences

Definition 5.6.1 We say that a term or qf formula of LR is *closed* if it has no variables in it. A closed qf formula is also called a *qf sentence*.

More generally, a formula of LR will be called a *sentence* if it has no free occurrences of variables in it. We keep this definition in reserve, to be used in Chapter 7; but note that qf formulas of LR contain no expressions that can bind variables, so all occurrences of variables in a qf formula are automatically free.

Suppose σ is a signature, A is a σ-structure and t is a closed term of LR(σ). Then A gives a meaning to each symbol in t (except the punctuation), so that we can read t and work out what it means. In the previous section we read the closed terms $\bar{S}(\bar{0})$ and $\bar{S}(\bar{S}(\bar{S}(\bar{S}(\bar{0}))))$ of LR(σ_{arith}) using the σ_{arith}-structure $\mathbb{N}$, and we saw that each of them names some element of $\mathbb{N}$. This is the normal situation: the closed term t of LR(σ), interpreted by A, names an element of A. We call this element t_A, and now we will see how to calculate it.

The following definition defines t_A for each closed term t, by recursion on the complexity of t. Note that if t is closed, then all subterms of t are closed too, so the definition makes sense. Also the value assigned to a complex term t_A depends only on the head of t and the values $(t_i)_A$ for the immediate subterms t_i of A.

Definition 5.6.2 Let σ be a signature and A a σ-structure. Then for each closed term t of LR(σ) we define the element t_A of A by:

(a) if t is a constant symbol c, then $t_A = c_A$;

(b) if F is an n-ary function symbol of σ and $t_1, \dots, t_n$ are closed terms of LR(σ), then $F(t_1, \dots, t_n)_A = F_A((t_1)_A, \dots, (t_n)_A)$.

(Here c_A and F_A are given by A as in Definition 5.5.2.) The value of t_A can be calculated compositionally by working up the parsing tree for t; see Exercise 5.6.3.

For example, in $\mathbb{N}$ we defined an element $\bar{0}_{\mathbb{N}}$ and functions $\bar{S}_{\mathbb{N}}$, $\bar{+}_{\mathbb{N}}$ and $\bar{\cdot}_{\mathbb{N}}$. With their help and Definition 5.6.2 we have

$$\begin{aligned}
\bar{0}_{\mathbb{N}} &= 0, \\
\bar{S}(\bar{0})_{\mathbb{N}} &= \bar{S}_{\mathbb{N}}(\bar{0}_{\mathbb{N}}) &= 1, \\
\bar{S}(\bar{S}(\bar{0}))_{\mathbb{N}} &= \bar{S}_{\mathbb{N}}(\bar{S}(\bar{0})_{\mathbb{N}}) &= 2, \\
\bar{\cdot}(\bar{S}(\bar{S}(\bar{0})), \bar{S}(\bar{S}(\bar{0})))_{\mathbb{N}} &= \bar{\cdot}_{\mathbb{N}}(2,2) &= 4
\end{aligned}$$

and so on. The next theorem is also an application of Definition 5.6.2. Note the notation $\bar{n}$, which we will use constantly when discussing $\mathbb{N}$.

Theorem 5.6.3 *Write $\bar{n}$ for the term*

$$\bar{S}(\bar{S}(\ldots \bar{S}(\bar{0})\ldots))$$

of $LR(\sigma_{arith})$ with exactly n $\bar{S}$'s. Then $\bar{n}_{\mathbb{N}} = n$.

Proof We use induction on n.
Case 1: $n = 0$. Then $\bar{n}$ is $\bar{0}$, and we know that $\bar{0}_{\mathbb{N}} = 0$.
Case 2: $n = k + 1$, assuming the result holds when $n = k$. We have:

$$\overline{k+1}_{\mathbb{N}} = \bar{S}(\bar{k})_{\mathbb{N}} = \bar{S}_{\mathbb{N}}(\bar{k}_{\mathbb{N}}) = k + 1,$$

where the middle equation is by Definition 5.6.2. $\square$

We turn now to qf sentences. If ϕ is a qf sentence of $LR(\sigma)$ and A is a σ-structure, then A interprets the symbols in ϕ so that ϕ becomes a true or false statement about A. In Chapter 3 we wrote $A^\star(\phi)$ for the truth value of ϕ in A, where ϕ was a formula of LP. But from now on we write

$$\models_A \phi$$

to mean that A makes ϕ true, and we read it as 'A is a model of ϕ'.

Definition 5.6.4 Let σ be a signature, A a σ-structure and χ a qf sentence of $LR(\sigma)$. Then the following definition (by recursion on the complexity of χ) tells us when '$\models_A \chi$' holds.

(a) If χ is $R(t_1, \ldots, t_n)$, where R is an n-ary relation symbol of σ and $t_1, \ldots, t_n$ are terms (necessarily closed since χ is closed), then

$$\models_A \chi \text{ if and only if } R_A((t_1)_A, \ldots, (t_n)_A).$$

(Here '$R_A((t_1)_A, \ldots, (t_n)_A)$' means that the n-tuple $((t_1)_A, \ldots, (t_n)_A)$ is in the relation R_A mentioned in Definition 5.5.2.)

(b) If χ is $(s = t)$ where s and t are terms (again necessarily closed), then

$$\models_A \chi \text{ if and only if } s_A = t_A.$$

(c) If χ is $\bot$ then $\models_A \chi$ does not hold (i.e. χ is false in A).

(d) If χ is $(\neg\phi)$ then

$$\models_A \chi \text{ if and only if it is not true that } \models_A \phi.$$

(e) If χ is $(\phi \wedge \psi)$ then

$$\models_A \chi \text{ if and only if both } \models_A \phi \text{ and } \models_A \psi.$$

(f)–(h) Similar clauses when χ is $(\phi \vee \psi)$, $(\phi \rightarrow \psi)$ or $(\phi \leftrightarrow \psi)$.

We say that A is a *model* of χ (or that χ is *true in* A) when $\models_A \chi$.

Remark 5.6.5 Since '$\models_A \phi$' is just another notation for '$A^\star(\phi) = T$', the definition above is the counterpart for qf LR of Definition 3.5.6 for LP. (You can reconstruct the missing clauses (f)–(h) above from their counterparts (e)–(g) in Definition 3.5.6.) We will keep the notation with $A^\star$ for the compositional definition of the truth value of a qf sentence in Appendix B, where it helps to give a clean and algebraic formulation. But the notation '$\models_A \chi$', though formally a little clumsier, seems closer to how most mathematicians think.

Example 5.6.6

(1) We know how to calculate that $\mp(\bar{1}, \dot{-}(\bar{3}, \bar{2}))_\mathbb{N} = 7$ and $\mp(\bar{3}, \bar{4})_\mathbb{N} = 7$. Hence

$$\models_\mathbb{N} \ (\mp(\bar{1}, \dot{-}(\bar{3}, \bar{2})) = \mp(\bar{3}, \bar{4})).$$

Usually we would write this as

$$\models_\mathbb{N} \ 1 + (3 \cdot 2) = 3 + 4.$$

(2) If ϕ is the formula $x_1 + x_2 = x_3 + Sx_4$ then we can turn it into a qf sentence by substituting closed terms for the variables. We have:

$$\models_\mathbb{N} \ \phi[2/x_1, 66/x_2, 67/x_3, 0/x_4],$$
$$\text{but not } \models_\mathbb{N} \ \phi[5/x_1, 6/x_2, 7/x_3, 8/x_4].$$

(3) For greater variety of examples, we introduce the signature σ_{oring}. ('oring' stands for 'ordered ring', but you will not need to know this.) At the same time we describe a particular σ_{oring}-structure, namely the ordered field of real numbers $\mathbb{R}$. Notice that we can save space this way: σ_{oring} can be inferred from the chart for $\mathbb{R}$. Its symbols are listed on the left, and their interpretations on the right show whether they are constant, function or relation symbols, and their arities.

The σ_{oring}-structure $\mathbb{R}$ is as follows:

(5.47)

domain	the set of all real numbers;
$\bar{0}$	the number 0;
$\bar{1}$	the number 1;
$\bar{-}(x)$	minus x;
$\bar{+}(x, y)$	$x + y$;
$\bar{\cdot}(x, y)$	xy;
$<(x, y)$	x is less than y.

The qf sentence

$$(-((1+1)+1) < (1+(1+(1+1))) \cdot (-1) \vee 1 = 1 + 0)$$

says that

$$-3 < -4 \quad \text{or} \quad 1 = 1,$$

which is true in $\mathbb{R}$. So

$$\models_{\mathbb{R}} \ -((1+1)+1) < (1+(1+(1+1))) \cdot (-1) \vee 1 = 1 + 0.$$

Just as in Section 3.5, we can use the definition of models to introduce several other important notions.

Definition 5.6.7 Let σ be a signature and qf ϕ a sentence of $\text{LR}(\sigma)$.

(a) We say that ϕ is *valid*, in symbols $\models_\sigma \phi$, if every σ-structure is a model of ϕ. (The word 'tautology' is used only for valid formulas of LP.)

(b) We say that ϕ is *consistent*, and that it is *satisfiable*, if some σ-structure is a model of ϕ.

(c) We say that ϕ is a *contradiction*, and that it is *inconsistent*, if no σ-structure is a model of ϕ.

(d) If Γ is a set of qf sentences of $\text{LR}(\sigma)$, then we say that a σ-structure A is a *model* of Γ if A is a model of every sentence in Γ. We write

(5.48) $$\Gamma \models_\sigma \phi$$

to mean that every σ-structure that is a model of Γ is also a model of ϕ. We call (5.48) a *semantic sequent*.

Exercise 5.6.4 will invite you to prove that, just as with LP, the subscript $_\sigma$ in $(\Gamma \models_\sigma \phi)$ makes no difference as long as ϕ and the formulas in Γ are all in $\text{LR}(\sigma)$.

The following fundamental lemma tells us that if a qf sentence ϕ contains a closed term t, then the only contribution that t makes to the truth or otherwise of ϕ in a structure A is the element t_A. If we replaced t in ϕ by a closed term s with $s_A = t_A$, then this would make no difference to the truth value in A. The reason for this is the compositional definition of semantics: in the definition of '$\models_A \phi$', t contributes no information beyond t_A.

Lemma 5.6.8 *Let σ be a signature and A a σ-structure. Let $t_1, \ldots, t_n$ and $s_1, \ldots, s_n$ be closed terms of $LR(\sigma)$ such that for each i ($1 \leqslant i \leqslant n$), $(s_i)_A = (t_i)_A$.*

(a) Let t be a term of $LR(\sigma)$ with distinct variables $x_1, \ldots, x_n$. Then

$$t[t_1/x_1, \ldots, t_n/x_n]_A = t[s_1/x_1, \ldots, s_n/x_n]_A.$$

(b) Let ϕ be a qf formula of $LR(\sigma)$ with distinct variables $x_1, \ldots, x_n$. Then

$$\models_A \phi[t_1/x_1, \ldots, t_n/x_n] \text{ if and only if } \models_A \phi[s_1/x_1, \ldots, s_n/x_n].$$

Proof We write T for the substitution $t_1/x_1, \ldots, t_n/x_n$ and S for the substitution $s_1/x_1, \ldots, s_n/x_n$. The proof of (a) is by induction on the complexity of the term t.

Case 1: t has complexity 0. There are three subcases.

(i) t is a constant symbol c. Then $t[T]_A = c_A = t[S]_A$.

(ii) t is x_i with $1 \leqslant i \leqslant n$. Then $t[T]_A = (t_i)_A = (s_i)_A = t[S]_A$.

(iii) t is a variable not among $x_1, \ldots, x_n$. Then $t[T]_A = t_A = t[S]_A$.

Case 2: t has complexity $k > 0$, where (a) holds for all terms of complexity $< k$. Then t is $F(r_1, \ldots, r_m)$ for some terms $r_1, \ldots, r_m$ all of complexity $< k$. We have

$$
\begin{aligned}
t[T]_A &= F(r_1[T], \ldots, r_m[T])_A = F_A(r_1[T]_A, \ldots, r_m[T]_A) \\
&= F_A(r_1[S]_A, \ldots, r_m[S]_A) = \ldots = t[S]_A
\end{aligned}
$$

where the first equation is by Definition 5.4.4(c), the second by Definition 5.6.2(b) and the third by induction hypothesis.

The proof of (b) is similar, by induction on the complexity of ϕ. $\square$

Exercises

5.6.1.   Write suitable clauses (f)–(h) for Definition 5.6.4. [See Remark 5.6.5.]

5.6.2.   Which of the following are true and which are false? (In (b)–(d) we allow ourselves names $\overline{m}$ for numbers m.)

     (a) $\models_{\mathbb{N}} \ (\cdot(\overline{4}, \overline{3}) = \bar{S}(\bar{S}(\dot{-}(\overline{2}, \overline{5}))))$.

     (b) $\models_{\mathbb{N}/5\mathbb{N}} \ \overline{4} \dot{+} \overline{4} = \overline{2}$.

(c) $\models_{\mathbb{R}} \overline{7} \cdot \bar{\pi} = \overline{22} \rightarrow \overline{8} \cdot \bar{\pi} = \overline{23}$.

(d) $\models_{\mathbb{R}} \overline{9} < \bar{\pi} \cdot \bar{\pi} \wedge \bar{\pi} \cdot \bar{\pi} < \overline{10}$.

5.6.3. Write a compositional definition of t_A for closed terms t of LR(σ) and a σ-structure A.

5.6.4. (a) (A version of the Principle of Irrelevance for qf LR) Suppose ρ and σ are signatures with $\rho \subseteq \sigma$, A is a σ-structure and B is a ρ-structure with the same domain as A, and $s_A = s_B$ for every symbol $s \in \rho$. Show that if ϕ is a qf sentence of LR(ρ), then $\models_A \phi$ if and only if $\models_B \phi$. [Adapt the proof of Lemma 5.6.8, starting with closed terms.]

(b) Using (a), show that if ρ and σ are signatures with $\rho \subseteq \sigma$, Γ is a set of qf sentences of LR(σ) and ψ is a qf sentence of LR(σ), then $(\Gamma \models_\rho \psi)$ holds if and only if $(\Gamma \models_\sigma \psi)$ holds. [This is the analogue for qf LR of Exercise 3.5.4, and the strategy of the proof is the same.]

5.6.5. Let ϕ be a qf sentence of LR(σ) which uses all the symbols in σ, and ℓ the length of ϕ (i.e. the number of symbols in ϕ), such that $\models_A \phi$ for some σ-structure A. Build a σ-structure B with at most ℓ elements, which is also a model of ϕ. [Let T be the set of all closed terms that occur in ϕ, including subterms of other terms. Then T has $< \ell$ members. Let W be the set of elements of A that are of the form t_A for some $t \in T$; so again W has $< \ell$ members. By Definition 5.5.2(a) the domain of A is not empty; choose an element a of A. The σ-structure B has domain $W \cup \{a\}$. The constant symbols are interpreted in B the same way as in A. The function symbols F are also interpreted the same way, except that if $b_1, \ldots, b_n$ are in $W \cup \{a\}$ and $F_A(b_1, \ldots, b_n)$ is not in $W \cup \{a\}$, then put $F_B(b_1, \ldots, b_n) = a$. You can work out what to do with the relation symbols, and how to show that this construction answers the question.]

5.7 Satisfaction

We say that the natural number 4 *satisfies* the condition $2 < x < 6$. One way of explaining this is as follows: if we take the name $\bar{4}$ of 4 and write it in the formula $(2 < x \wedge x < 6)$ in place of x, we get a statement $(2 < \bar{4} \wedge \bar{4} < 6)$ which is true.

Example 5.7.1 Suppose ϕ is the formula

$$(\neg(\bar{S}(x) = \bar{S}(\bar{S}(\bar{0}))))$$

of $\text{LR}(\sigma_{\text{arith}})$, or in shorthand

$$S(x) \neq \bar{2}$$

Then it expresses the condition that $x + 1 \neq 2$, in other words $x \neq 1$. The numbers 0 and 2 satisfy ϕ in $\mathbb{N}$, but the number 1 fails to satisfy it. Likewise:

- Every natural number satisfies $(x = x)$ in $\mathbb{N}$.
- No natural number satisfies $(\bot \wedge (x = x))$ in $\mathbb{N}$.
- 0 and 1 are the only numbers satisfying $(x \cdot x = x)$ in $\mathbb{N}$.
- 0 and 2 are the only numbers not satisfying $(\neg(x \cdot x = x + x))$ in $\mathbb{N}$.

We would like to generalise this idea to n-tuples of elements. For example, it would be useful to be able to say that the ordered pair $(3.5, 5.5)$ satisfies the condition $x < y$ in $\mathbb{R}$, but the ordered pair $(88, 42)$ does not. But we run into two problems.

Problem 5.7.2 First, how do we tell which item in the pair goes with which variable? In the case of $(3.5, 5.5)$ and $x < y$ it is a reasonable guess that 3.5 goes with x and 5.5 goes with y. But, for example, this is no help for deciding whether the pair $(3.5, 5.5)$ satisfies $z < 4$; does z pick up 3.5 or 5.5?

In mathematical practice the same problem arises with functions of more than one variable, and the usual solution there will work for us too.

Definition 5.7.3 Let $y_1, \ldots, y_n$ be distinct variables. If we introduce a formula ϕ as $\phi(y_1, \ldots, y_n)$, this means that when we apply an n-tuple $(a_1, \ldots, a_n)$ of elements to ϕ, we attach a_1 to y_1, a_2 to y_2 and so on. (The listed variables should include all the variables free in ϕ, though sometimes it is useful to include some other variables as well.)

So, for example, if ϕ is $x + y = z$, then $(2, 3, 5)$ satisfies $\phi(x, y, z)$ and $\phi(y, x, z)$ in $\mathbb{N}$. It does not satisfy $\phi(z, y, x)$, but $(5, 3, 2)$ does. Also $(2, 3, 5, 1)$ satisfies $\phi(x, y, z, w)$; in fact, $(2, 3, 5, n)$ satisfies $\phi(x, y, z, w)$ regardless of what natural number n is.

Problem 5.7.4 Second, our explanation of satisfaction assumed that we have closed terms naming the elements in question; we substitute these names for the variables and then see whether the resulting sentence is true. This works in the structure $\mathbb{N}$ because every natural number n has a name $\bar{n}$ in $\text{LR}(\sigma_{\text{arith}})$. But, for example, the signature σ_{digraph} has no constant symbols and hence no closed terms. So if G is a digraph, no element at all of G is named by a closed term of $\text{LR}(\sigma_{\text{digraph}})$. But we surely want to be able to ask whether an element of a digraph satisfies a formula.

The next definition shows a way of doing it: we choose a new symbol and add it to the signature as a name of the element.

Definition 5.7.5 Let σ be a signature. A *witness* is a new constant symbol that is added to σ. We use the notation $\bar{a}$ for a witness added to name the element a in a σ-structure A; so $\bar{a}_A = a$.

Definition 5.7.6 Suppose $\phi(y_1, \ldots, y_n)$ is a qf formula of $\mathrm{LR}(\sigma)$, A is a σ-structure and $a_1, \ldots, a_n$ are elements in the domain of A. We define the relation '$(a_1, \ldots, a_n)$ *satisfies* ϕ in A' by the following recursion on the complexity of ϕ:

(a) If ϕ is atomic then $(a_1, \ldots, a_n)$ satisfies ϕ in A if and only if

$$(5.49) \qquad \models_A \phi[t_1/y_1, \ldots, t_n/y_n]$$

where $t_1, \ldots, t_n$ are closed terms (possibly witnesses) such that for each i, $(t_i)_A = a_i$. (For this definition to make sense, we have to know that the truth of (5.49) depends only on $a_1, \ldots, a_n$ and not on the choice of terms used to name them. But Lemma 5.6.8(b) guarantees this.)

(b) If ϕ is $(\neg\psi)$, then $(a_1, \ldots, a_n)$ satisfies ϕ if and only if $(a_1, \ldots, a_n)$ does not satisfy $\psi(y_1, \ldots, y_n)$.

(c) If ϕ is $(\psi \wedge \chi)$, then $(a_1, \ldots, a_n)$ satisfies ϕ if and only if $(a_1, \ldots, a_n)$ satisfies both $\psi(y_1, \ldots, y_n)$ and $\chi(y_1, \ldots, y_n)$.

(d) If ϕ is $(\psi \leftrightarrow \chi)$, then $(a_1, \ldots, a_n)$ satisfies ϕ if and only if either $(a_1, \ldots, a_n)$ satisfies both $\psi(y_1, \ldots, y_n)$ and $\chi(y_1, \ldots, y_n)$, or it satisfies neither.

(e), (f) Similar clauses for $(\psi \vee \chi)$ and $(\psi \rightarrow \chi)$, tracking the truth tables for $\vee$ and $\rightarrow$.

Consider, for example, the real numbers in the form of the σ_{oring}-structure $\mathbb{R}$, and let $\phi(x, y)$ be the formula $(y < x)$. To check whether the pair (e, π) satisfies ϕ in $\mathbb{R}$, we ask whether the sentence

$$(\bar{\pi} < \bar{e})$$

is true in $\mathbb{R}$ (where $\bar{e}$ and $\bar{\pi}$ are witnesses naming the transcendental numbers e and π). Since $e = 2.7\ldots$ and $\pi = 3.1\ldots$, the answer is 'No,' and hence (e, π) does not satisfy the formula in $\mathbb{R}$. But the condition for $\neg$ in Definition 5.7.6(b) tells us that (e, π) does satisfy the formula $(\neg\phi)$.

Definition 5.7.7 Let $\phi(y_1, \ldots, y_n)$ and $\psi(y_1, \ldots, y_n)$ be qf formulas of $LR(\sigma)$ and A a σ-structure. We say that ϕ and ψ are *equivalent in A* if

> for all elements $a_1, \ldots, a_n$ of A, $(a_1, \ldots, a_n)$ satisfies ϕ in A if and only if it satisfies ψ in A.

We say that ϕ and ψ are *logically equivalent* if they are equivalent in every σ-structure.

When we extend this definition to formulas with quantifiers in Section 7.3, we will find various important logical equivalences. But for the moment we note one obvious example.

Lemma 5.7.8 *Let s and t be any terms of $LR(\sigma)$. Then $(s = t)$ and $(t = s)$ are logically equivalent formulas.*

Proof Let $y_1, \ldots, y_n$ be all the variables that occur in s or t, listed without repetition. Let $\phi(y_1, \ldots, y_n)$ be $(s = t)$ and let $\psi(y_1, \ldots, y_n)$ be $(t = s)$. Let A be a σ-structure and $a_1, \ldots, a_n$ elements of A. We must show that $(a_1, \ldots, a_n)$ satisfies ϕ if and only if it satisfies ψ. Take witnesses $\bar{a}_1$, etc. to name the elements a_1, etc.

Now $\phi[\bar{a}_1/y_1, \ldots, \bar{a}_n/y_n]$ is the result of putting $\bar{a}_1$ for y_1 etc. everywhere through $(s = t)$; so it is a sentence of the form $(s' = t')$ for some closed terms s' and t'. Since the same substitutions are made in the formula $\psi[\bar{a}_1/y_1, \ldots, \bar{a}_n/y_n]$, this formula is the sentence $(t' = s')$. But

$$\models_A (s' = t') \quad \Leftrightarrow \quad (s')_A = (t')_A \quad \text{by Definition 5.6.4(b)}$$
$$\Leftrightarrow \quad (t')_A = (s')_A$$
$$\Leftrightarrow \quad \models_A (t' = s') \quad \text{by Definition 5.6.4(b)} \qquad \Box$$

We remark that in the statement '$\phi(x_1, \ldots, x_n)$ and $\psi(x_1, \ldots, x_n)$ are equivalent in A', the listing of the variables is not really needed. To test logical equivalence, one needs to make sure that the same term is put for x_1 in both ϕ and ψ, and similarly for the other variables; but for this the order does not matter. The same applies to logical equivalence. However, in the proof of Lemma 5.7.8 it was convenient to have a listing of the variables.

Exercises

5.7.1. For each of the following formulas, say what is the set of natural numbers that satisfy the formula in $\mathbb{N}$.

 (a) $(x = x \dotplus x)$.

(b) $(\bar{S}(x) = x\dot{-}x)$.

(c) $(\neg((x = \bar{7}) \lor (x = \bar{9})))$.

(d) $(x\dot{-}\bar{S}(x) = x\dot{+}\overline{25})$.

5.7.2. For each of the following formulas $\phi(x, y)$, say what is the set of pairs (m, n) of natural numbers that satisfy the formula in $\mathbb{N}$.

(a) $(x\dot{-}y = \bar{0})$.

(b) $((x = \bar{4}) \land (y = \overline{1066}))$.

(c) $(x = \bar{4})$.

(d) $(x = \bar{S}(y))$.

(e) $(x\dot{-}\bar{S}(y) = \bar{S}(x)\dot{-}y)$.

(f) $((x\dot{-}x)\dot{-}(y\dot{-}y) = \overline{36})$.

5.7.3. In each of the following cases, describe the set of pairs (a, b) of elements that satisfy ϕ in the given structure. In this exercise we use ordinary mathematical notation, but the formulas can be written as qf formulas of LR.

(a) $\phi(x, y)$ is $x^2 + y^2 = 1$, the structure is $\mathbb{R}$.

(b) $\phi(x, y)$ is $x^2 + y^2 < 1$, the structure is $\mathbb{R}$.

(c) $\phi(x, y)$ is $y < 1$, the structure is $\mathbb{R}$.

(d) $\phi(x, y)$ is $x + y = 1$, the structure is $\mathbb{N}/5\mathbb{N}$.

(e) $\phi(x, y)$ is $x^2 + y^2 = 4$, the structure is $\mathbb{N}/5\mathbb{N}$.

(f) $\phi(x, y)$ is $x^2 y^2 = 2$, the structure is $\mathbb{N}/7\mathbb{N}$.

5.7.4. Show that the two formulas $(x = y)$ and $(\bar{S}(x) = \bar{S}(y))$ are equivalent in $\mathbb{N}$.

5.7.5. Show that the two formulas $((x = \bar{0}) \land (y = \bar{0}))$ and $(x^2 + y^2 = \bar{0})$ are equivalent in $\mathbb{R}$.

5.7.6. Let $\phi(x)$ be a qf formula of $\mathrm{LR}(\sigma_{\mathrm{arith}})$. Show that the set of natural numbers that satisfy ϕ in $\mathbb{N}$ consists of either finitely many numbers, or all the natural numbers except finitely many. [This asks for an induction on the complexity of ϕ.]

5.8 Diophantine sets and relations

Diophantine relations provide some excellent insight into the notion of satisfaction. This is one reason why we explore them here. But later we will discover some powerful facts about them that are directly relevant to Leibniz's project of doing logic by calculation. In the next definition, remember from Definition

5.2.3 that an n-ary relation on a set X is the same thing as a set of n-tuples of elements of X.

Definition 5.8.1 Let n be a positive integer and X a set of n-tuples of natural numbers. We say that X is *diophantine* if

there is an equation $\phi(x_1, \ldots, x_n, y_1, \ldots, y_m)$ of $\mathrm{LR}(\sigma_{\mathrm{arith}})$ such that for every n-tuple $(k_1, \ldots, k_n)$ of natural numbers,

$$(k_1, \ldots, k_n) \in X \quad \Leftrightarrow \quad \text{for some numbers } \ell_1, \ldots, \ell_m,$$
$$\models_{\mathbb{N}} \phi(\bar{k}_1, \ldots, \bar{k}_n, \bar{\ell}_1, \ldots, \bar{\ell}_m).$$

For example, the set of natural numbers that can be written as the sum of three squares is the set of natural numbers k such that

(5.50) $$\text{for some } \ell_1, \ell_2, \ell_3, \quad k = \ell_1^2 + \ell_2^2 + \ell_3^2$$

Consider the equation

(5.51) $$x = y_1^2 + y_2^2 + y_3^2$$

We can write (5.51) as the following equation $\phi(x, y_1, y_2, y_3)$ of $\mathrm{LR}(\sigma_{\mathrm{arith}})$:

(5.52) $$(x = \mp(\mp(\bar{\cdot}(y_1, y_1), \bar{\cdot}(y_2, y_2)), \bar{\cdot}(y_3, y_3)))$$

Then for every k,

$$k \text{ is the sum of three squares} \quad \Leftrightarrow \quad \text{for some numbers } \ell_1, \ell_2, \ell_3,$$
$$\models_{\mathbb{N}} \phi(\bar{k}, \bar{\ell}_1, \bar{\ell}_2, \bar{\ell}_3).$$

So the set of sums of three squares is diophantine.

As this example illustrates, a direct proof that a relation is diophantine involves finding an equation of $\mathrm{LR}(\sigma_{\mathrm{arith}})$. But by now you should know which equations in ordinary mathematical notation can be rewritten as equations of $\mathrm{LR}(\sigma_{\mathrm{arith}})$, so we will use the ordinary notation.

Theorem 5.8.2 *The following sets and relations on the natural numbers are diophantine.*

(a) x is even.

(b) x is a cube.

(c) $x_1 \leqslant x_2$.

(d) $x_1 < x_2$.

(e) x_1 *divides* x_2.

(f) $x_1 + 1, x_2 + 1$ *are relatively prime.*

(g) $x_1 \neq x_2$.

(h) x *is not a power of 2.*

Proof (a) x is even if and only if for some natural number y, $x = 2y$.

(c) $x_1 \leqslant x_2$ if and only if for some natural number y, $x_1 + y = x_2$.
The rest are exercises. ☐

Theorem 5.8.3 *Let* $s_1 = t_1$ *and* $s_2 = t_2$ *be two equations in* $LR(\sigma_{arith})$. *Then*

(1) the formula $s_1 = t_1 \wedge s_2 = t_2$ *is equivalent in* $\mathbb{N}$ *to an equation;*

(2) the formula $s_1 = t_2 \vee s_2 = t_2$ *is equivalent in* $\mathbb{N}$ *to an equation.*

The proof will use the following device. Suppose E is an equation in the *integers*, and one of the terms of E has the form $-t$. Then by adding t to both sides and cancelling, we can remove the occurrence of $-$. The medieval Arab mathematicians referred to this operation as *al-jabr*. By reducing both sides of E to simplest terms and then applying al-jabr enough times, we can reach an equation E' that is equivalent to E in the integers and has no occurrences of $-$. The operation is mathematically trivial, but our reason for applying it is not: in order to prove a fact about $\mathbb{N}$ we pass to $\mathbb{Z}$ (the integers), and then we need a device for getting back to $\mathbb{N}$.

(The name 'al-jabr' originally meant applying force. It became specialised to the application of force to a broken or dislocated bone to push it back into place. Then the mathematicians took it over, with the idea that a term $-t$ has got itself dislocated onto the wrong side of an equation. The word then evolved from 'al-jabr' into 'algebra'. Another word from the same root is the name Gabriel, 'God is my force'.)

Proof (a) We have the equivalences

$$
\begin{array}{ll}
& s_1 = t_1 \wedge s_2 = t_2 & \text{(in } \mathbb{N}） \\
\Leftrightarrow & (s_1 - t_1) = 0 = (s_2 - t_2) & \text{(in } \mathbb{Z}） \\
\Leftrightarrow & (s_1 - t_1)^2 + (s_2 - t_2)^2 = 0 & \text{(in } \mathbb{Z}） \\
\Leftrightarrow & s_1^2 + t_1^2 + s_2^2 + t_2^2 = 2s_1 t_1 + 2s_2 t_2 & \text{(in } \mathbb{N} \text{ by al-jabr)}
\end{array}
$$

(b) Likewise

$$
\begin{array}{ll}
& s_1 = t_1 \vee s_2 = t_2 & \text{(in } \mathbb{N}） \\
\Leftrightarrow & (s_1 - t_1) = 0 \text{ or } (s_2 - t_2) = 0 & \text{(in } \mathbb{Z}） \\
\Leftrightarrow & (s_1 - t_1)(s_2 - t_2) = 0 & \text{(in } \mathbb{Z}） \\
\Leftrightarrow & s_1 s_2 + t_1 t_2 = s_1 t_2 + t_1 s_2 & \text{(in } \mathbb{N} \text{ by al-jabr)} \quad ☐
\end{array}
$$

So, for example, the set of natural numbers n that are *not* prime is diophantine:

$$
\begin{aligned}
n \text{ not prime} \quad\Leftrightarrow\quad & \text{for some } y, z, \\
& (n = 0) \vee (n = 1) \vee ((y < n) \wedge (z < n) \wedge (n = yz)) \\
\Leftrightarrow\quad & \text{for some } y, z, u, v, \\
& (n = 0) \vee (n = 1) \vee \\
& ((n = y + Su) \wedge (n = z + Sv) \wedge (n = yz)).
\end{aligned}
$$

We turn to an interesting description of the diophantine sets.

Definition 5.8.4 Let n be a positive integer and let X be a set of n-tuples $(k_1, \dots, k_m)$ of natural numbers.

We say that X is *computably enumerable*, or *c.e.* for short, if a digital computer with infinitely extendable memory and unlimited time can be programmed so that it lists (not necessarily in any reasonable order) all and only the n-tuples that are in X. (Imagine it programmed so that every time you press the Return key, it prints out another n-tuple from the list.)

We say that X is *computable* if a digital computer with infinitely extendable memory and unlimited time can be programmed so that if you type in an n-tuple in X, the computer prints YES, and if you type in an n-tuple not in X then it prints NO.

Computably enumerable sets used to be known as *recursively enumerable*, or *r.e.* for short. Likewise computable sets have often been called *recursive*. There is a move to replace these old names, because being computable or c.e. has little to do with recursion and everything to do with computation.

We write $\mathbb{N}^n$ for the set of all n-tuples of natural numbers. We write $A \setminus B$ for the set of elements of A that are not in B.

Lemma 5.8.5 *Let n be a positive integer and X a set of n-tuples of natural numbers. Then the following are equivalent:*

(a) X is computable.

(b) Both X and $\mathbb{N}^n \setminus X$ are c.e.

Proof (a) $\Rightarrow$ (b): Assuming (a), we can program a computer so that it runs through all n-tuples of natural numbers in turn, tests each one to see if it is in X, then prints out the n-tuple when the answer is YES. This shows that X is c.e.; the same argument applies to $\mathbb{N}^n \setminus X$, using the answer NO.

(b) $\Rightarrow$ (a): Assume X and $\mathbb{N}^n \setminus X$ are both c.e. Then we can program a computer to test any n-tuple K for membership of X as follows. The computer simultaneously makes two lists, one of X and one of $\mathbb{N}^n \setminus X$, until K turns up in one of the lists (as it must eventually do). If the list that K appears in is that of X, the computer prints YES; if the list is of $\mathbb{N}^n \setminus X$ then it prints NO. □

Since at least one of X and $\mathbb{N}^n \setminus X$ is infinite, we have no guarantee that K is going to turn up in either list before the human race becomes extinct. So this is not a helpful computer program. But it does give the right answer.

Yuri Matiyasevich Russia, living.
There is no algorithm for telling whether a diophantine equation has a solution.

Theorem 5.8.6 (Matiyasevich's Theorem) *Let n be a positive integer and X a set of n-tuples of natural numbers. Then the following are equivalent:*

(a) X is computably enumerable.

(b) X is diophantine.

Proof (b) ⇒ (a) is clear: we program a computer to test all $(n + m)$-tuples of natural numbers in turn, and whenever it reaches one such that

$$(5.53) \qquad \models_{\mathbb{N}} \phi(\bar{k}_1, \dots, \bar{k}_n, \bar{\ell}_1, \dots, \bar{\ell}_m),$$

it prints out $(k_1, \dots, k_n)$. How does the computer check whether (5.53) holds? Since ϕ is an equation in natural numbers, the computer only has to calculate the values of the left side and the right side and then check that they are equal.

The proof of direction (a) ⇒ (b) is difficult and we omit it. The idea is that an n-tuple K is in X if and only if there is a computation C according to a given computer program, that returns the answer YES when we type in K. Using Alan Turing's analysis of computing, we can code up the computation C as a set of numbers satisfying certain conditions. To express these conditions by a natural number equation involves a deep study of diophantine relations. (For further information see Yuri V. Matiyasevich, *Hilbert's Tenth Problem*, MIT Press, Cambridge, MA, 1993.) □

Exercises

5.8.1. Show that the empty set of natural numbers and the set of all natural numbers are both diophantine.

5.8.2. Prove the remaining cases of Theorem 5.8.2.

5.8.3. Show that the following relations are diophantine.

  (a) x_3 is the remainder when x_1 is divided by $x_2 + 1$.

  (b) x_3 is the integer part of the result of dividing x_1 by $x_2 + 1$.

5.8.4. Show that if n is a positive integer and R, S are diophantine n-ary relations on the natural numbers, then $R \cap S$ and $R \cup S$ are also diophantine. [Use Theorem 5.8.3.]

5.8.5. The two following statements are equivalent:

  (i) $F(G(x_1)) = x_2$.

  (ii)   There is y such that $G(x_1) = y$ and $F(y) = x_2$.

  After you have satisfied yourself that this is true, use this idea (and the proof of Theorem 5.8.3) to show that every diophantine relation can be written as in Definition 5.8.1 using an equation ϕ of the form $s = t$ in which both s and t are polynomials in $x_1, \dots, y_1, \dots$ of degree $\leqslant 4$.

5.9 Soundness for qf sentences

Suppose Γ is a set of qf formulas of $\mathrm{LR}(\sigma)$ and ψ is a qf formula of $\mathrm{LR}(\sigma)$. In Definition 5.4.7 we defined what

$$(5.54) \qquad\qquad\qquad \Gamma \vdash_\sigma \psi$$

means, in terms of derivations. In Definition 5.6.7(d) we defined what

$$(5.55) \qquad\qquad\qquad \Gamma \models_\sigma \psi$$

means, in terms of models.

Theorem 5.9.1 (Soundness of Natural Deduction with qf Sentences) *Let σ be a signature, Γ a set of qf sentences of $LR(\sigma)$ and ψ a qf sentence of $LR(\sigma)$. If $\Gamma \vdash_\sigma \psi$ then $\Gamma \models_\sigma \psi$.*

Proof   We show that

(5.56)   If D is a σ-derivation whose conclusion is ψ and whose undischarged assumptions all lie in Γ, and which uses no formulas except qf sentences of $\mathrm{LR}(\sigma)$, then every σ-structure that is a model of Γ is also a model of ψ.

As in the proof of Soundness for propositional logic (Theorem 3.9.2) we use the fact that D is a tree; so we can prove (5.56) for all Γ, ψ and D by induction on the height of D.

The cases covered in the proof of Theorem 3.9.2 go exactly as before, but using Definition 5.4.5 of σ-derivations. There remain the two cases where the right label R on the bottom node of D is either $(= I)$ or $(= E)$. We assume D has height k and (5.56) is true for all derivations of height $< k$ (and all Γ and ψ).

Case 2 (d): R is $(=I)$. In this case ψ has the form $(t = t)$ where t is a closed term (because it is in a qf sentence). Let A be any σ-structure. Then by Definition 5.6.4(b),

$$(5.57) \qquad \models_A (t = t) \;\Leftrightarrow\; t_A = t_A$$

But t_A is trivially equal to t_A, so $\models_A (t = t)$ as required.

Case 2 (e): R is $(=E)$. Then D has the form

$$\begin{array}{cc} D_1 & D_2 \\ (s = t) & \phi[s/x] \\ \hline \multicolumn{2}{c}{\phi[t/x]} \end{array}$$

No assumptions are discharged here, so that both D_1 and D_2 are derivations whose undischarged assumptions all lie in Γ. They have height $< k$, so by induction hypothesis, every model of Γ is also a model of $(s = t)$ and $\phi[s/x]$. Thus if A is any model of Γ then $s_A = t_A$ by the first, and

$$\models_A \phi[s/x]$$

by the second. Since $s_A = t_A$, Lemma 5.6.8(b) guarantees that

$$\models_A \phi[t/x]$$

as well. But $\phi[t/x]$ is ψ, as required.

So (5.56) is proved. This is not quite the end of the story, because there could be a σ-derivation D that proves the sequent $\Gamma \vdash \psi$ and contains formulas that are not sentences. To deal with this case, let τ be a signature containing a constant symbol c, such that $\sigma \subseteq \tau$. (If σ already contains a constant symbol, take τ to be σ.) As in Exercise 5.4.7 we can put c in place of every variable, so as to get a τ-derivation that proves $\Gamma \vdash \psi$ and uses only sentences. So by the case already proved, $\Gamma \models_\tau \psi$. Then $\Gamma \models_\sigma \psi$ as in Exercise 5.6.4(b) (basically the Principle of Irrelevance). $\square$

5.10 Adequacy and completeness for qf sentences

As in Section 3.10, and for the same reason, we assume that the symbols of the signature σ can be listed as $S_0, S_1, \ldots$. We will also assume that the signature σ contains at least one constant symbol—but see the comment at the end of the section.

Theorem 5.10.1 (Adequacy of Natural Deduction for qf Sentences) *Let Γ be a set of qf sentences of $LR(\sigma)$ and ψ a qf sentence of $LR(\sigma)$. Then*

$$\Gamma \models_\sigma \psi \;\Rightarrow\; \Gamma \vdash_\sigma \psi.$$

As in Section 3.10, for simplicity our proof applies to a stripped-down version of LR that never uses $\vee$, $\to$ or $\leftrightarrow$. As in Lemma 3.10.3, it will suffice to show that if Γ is a syntactically consistent set (Definition 5.4.7), then Γ has a model. To show that this suffices, we previously used Example 3.4.3 for propositional logic, but the natural deduction rules used there are still valid for our expanded logic.

Definition 5.10.2 Let σ be a signature and let Δ be a set of qf sentences of $LR(\sigma)$. We say that Δ is a *Hintikka set* (for our stripped-down $LR(\sigma)$) if it meets the following conditions:

(1) If a formula $(\phi \wedge \psi)$ is in Δ then ϕ is in Δ and ψ is in Δ.

(2) If a formula $(\neg(\phi \wedge \psi))$ is in Δ then either $(\neg\phi)$ is in Δ or $(\neg\psi)$ is in Δ.

(3) If a formula $(\neg(\neg\phi))$ is in Δ then ϕ is in Δ.

(4) $\bot$ is not in Δ.

(5) There is no atomic sentence ϕ such that both ϕ and $(\neg\phi)$ are in Δ.

(6) For every closed term t of $LR(\sigma)$, $(t = t)$ is in Δ.

(7) If ϕ is atomic and s, t are closed terms such that both $(s = t)$ and $\phi[s/x]$ are in Δ, then $\phi[t/x]$ is in Δ.

Lemma 5.10.3 *If Δ is a Hintikka set for $LR(\sigma)$, then some σ-structure is a model of Δ.*

Proof Let Δ be a Hintikka set for $LR(\sigma)$. We write C for the set of closed terms of $LR(\sigma)$, and we define a relation $\sim$ on C by

(5.58) $s \sim t \;\Leftrightarrow\; (s = t) \in \Delta.$

We claim that $\sim$ is an equivalence relation on C.

The relation $\sim$ is reflexive by (6). It is symmetric by (6), (7) and the calculation in Example 5.4.2, and it is transitive by a similar calculation. So the claim is true. We write $t^\sim$ for the set of all the closed terms s in C such that $s \sim t$. (This set is the equivalence class of t for the equivalence relation $\sim$.)

Now we can describe the required σ-structure A. We take as typical symbols of σ a constant symbol c, a 1-ary function symbol F and a 2-ary relation symbol R. Then A is defined by the following chart.

(5.59)

domain	the set of all $t^\sim$ with t in C;
c	the element $c^\sim$;
$F_A(a)$	the element $F(t)^\sim$ where a is $t^\sim$;
$R_A(a_1, a_2)$	$R(t_1, t_2) \in \Delta$, where a_1 is $t_1^\sim$ and a_2 is $t_2^\sim$.

The domain is non-empty—as required by Definition 5.5.2—because of our assumption that σ has a constant term.

There is a pitfall here in the interpretation of the function symbol F. The interpretation tells us that if a is an element of A, then we can find the element $F_A(a)$ as follows:

(5.60) Find the closed term t such that a is $t^\sim$ (we know that all the elements of A have this form), write the term $F(t)$ and then take $F(t)^\sim$; this element of A is the required value $F_A(a)$.

If you do not see at once what the problem is, probably you should stop reading and spend some time getting it into focus while we show you a picture of Leopold Kronecker, who first proved a version of Lemma 5.10.3.

Leopold Kronecker Germany, 1823–1891.
To build a model of a set of equations, make the model out of the symbols in the equations.

The problem with (5.60) is that there are in general many closed terms t such that $t^\sim = a$. In fact every term in the equivalence class $t^\sim$ will do. To justify our interpretation of F, we have to show that it does not matter which t you choose in the equivalence class; all of them lead to the same $F(t)^\sim$. We show this as follows.

Suppose $s \sim t$. Then by (5.58),

(5.61) $$(s = t) \text{ is in } \Delta$$

Now by (6) we have

$$(F(s) = F(s)) \text{ is in } \Delta$$

So by (7) applied with ϕ the atomic formula $(F(s) = F(x))$, where x is a variable, and (5.61),

$$(F(s) = F(t)) \text{ is in } \Delta$$

Then by (5.58) again,

$$(5.62) \qquad\qquad\qquad F(s) \sim F(t)$$

so $F(s)^\sim = F(t)^\sim$ as required.

The same problem arises with the interpretation of R. We need to show that when a_1, a_2 are elements of A, then the question whether $R_A(a_1, a_2)$ holds does not depend on the choice of closed terms t_1 and t_2 with $a_1 = t_1^\sim$ and $a_2 = t_2^\sim$.

Suppose $s_1^\sim = t_1^\sim$ and $s_2^\sim = t_2^\sim$. We need to show that $R(s_1, s_2) \in \Delta$ if and only if $R(t_1, t_2) \in \Delta$. Assume $R(s_1, s_2) \in \Delta$. By (5.58), $(s_1 = t_1) \in \Delta$. Then by (7) with s_1, t_1 in place of s, t and ϕ equal to $R(x, s_2)$, where x is a variable, we see that $R(t_1, s_2) \in \Delta$. Also $(s_2 = t_2) \in \Delta$. By a second application of (7) with s_2, t_2 in place of s, t and ϕ equal to $R(t_1, x)$, it follows that $R(t_1, t_2) \in \Delta$.

Now a symmetrical argument, swapping s_1 and s_2 with t_1 and t_2, shows that if $R(t_1, t_2) \in \Delta$ then $R(s_1, s_2) \in \Delta$, as required.

In general, if F is a function symbol of arity n and $a_1, \ldots, a_n$ are elements of A, then we interpret F so that $F_A(a_1, \ldots, a_n) = F(t_1, \ldots, t_n)^\sim$, where t_i is a closed term with $t_i^\sim = a_i$ for $1 \le i \le n$. Likewise, if R is a relation symbol of arity n, then we define $R_A(a_1, \ldots, a_n)$ to be true in A if and only if $R(t_1, \ldots, t_n) \in \Delta$, where again $t_i^\sim = a_i$. Similar but more elaborate arguments show that these definitions do not depend on the choice of closed terms t_i such that $t_i^\sim = a_i$ (see Exercise 5.10.1).

So the σ-structure A is defined. It remains to show that every qf sentence in the Hintikka set Δ is true in A.

A key observation is that for every closed term t,

$$(5.63) \qquad\qquad\qquad t_A = t^\sim.$$

This is proved by induction on the complexity of t. If t is a constant symbol c then $c_A = c^\sim$ by definition. If t is $F(t_1, \ldots, t_r)$ then

$$
\begin{aligned}
t_A &= F_A((t_1)_A, \ldots, (t_r)_A) && \text{by Definition 5.6.2(b)} \\
&= F_A(t_1^\sim, \ldots, t_r^\sim) && \text{by induction hypothesis} \\
&= F(t_1, \ldots, t_r)^\sim = t^\sim && \text{by definition of } F_A.
\end{aligned}
$$

This proves (5.63).

Now we show for each qf sentence ϕ, by induction on the complexity of ϕ, that

(5.64)   (a) if ϕ is in Δ then A is a model of ϕ, and
          (b) if $\neg\phi$ is in Δ then A is not a model of ϕ.

Case 1: ϕ is an atomic sentence of the form $(s = t)$. If $(s = t) \in \Delta$ then $s \sim t$, so using (5.63),

$$s_A = s^\sim = t^\sim = t_A$$

and hence $\models_A (s = t)$ by Definition 5.6.4(b).

If $\neg(s = t)$ is in Δ then $(s = t)$ is not in Δ, by clause (5), so $s \not\sim t$ and Definition 5.6.4(b) again shows that A is not a model of $(s = t)$.

Case 2: ϕ is an atomic sentence of the form $R(t_1, \ldots, t_n)$. This is very similar to Case 1, but a little simpler.

Case 3: ϕ is $\perp$ or $(\neg\psi)$ or $(\psi \wedge \chi)$. These cases are exactly as in the proof of Lemma 3.10.5 for LP, except that now we write $\models_A \phi$ instead of $A^*(\phi) = T$.

Now by (5.64)(a) we have checked that A is a model of Δ, so the lemma is proved. $\square$

Lemma 5.10.4 *If Γ is a syntactically consistent set of qf sentences of $LR(\sigma)$, there is a Hintikka set Δ for $LR(\sigma)$, such that $\Gamma \subseteq \Delta$.*

Proof   The proof is similar to that of Lemma 3.10.6 for LP. We make a listing

$$\theta_0, \theta_1, \ldots$$

where each θ_i is

EITHER a qf sentence of $LR(\sigma)$ not containing $\rightarrow$, $\leftrightarrow$ or $\vee$,

OR a triple of the form $(\phi, x, s = t)$ where ϕ is an atomic formula of $LR(\sigma)$ with the one variable x, and s, t are closed terms of $LR(\sigma)$.

We arrange the list so that each of these sentences or triples occurs infinitely often.

We define a sequence of sets Γ_i of qf sentences, where $\Gamma_0 = \Gamma$ and the definition of each Γ_{i+1} depends on Γ_i and θ_i. Suppose Γ_i has just been defined.

- If θ_i is a sentence of one of the forms $(\psi \wedge \chi)$ or $\neg(\psi \wedge \chi)$ or $\neg\neg\psi$, then we proceed exactly as in the proof of Lemma 3.10.6.

- If θ_i is of the form $(t = t)$ then we put $\Gamma_{i+1} = \Gamma_i \cup \{(t = t)\}$.
- If θ_i is $(\phi, x, s = t)$ and the qf sentences $(s = t)$ and $\phi[s/x]$ are in Γ_i, then we put $\Gamma_{i+1} = \Gamma_i \cup \{\phi[t/x]\}$.
- In all other cases, $\Gamma_{i+1} = \Gamma_i$.

After all the Γ_i have been defined, we take Δ to be the union $\Gamma_0 \cup \Gamma_1 \cup \cdots$. We claim that Δ is syntactically consistent. As in the proof of Lemma 3.10.6, this reduces to showing by induction on i that for each i, if Γ_i is syntactically consistent then so is Γ_{i+1}.

Besides the cases (i), (ii) and (iii) in the proof of Lemma 3.10.6 there are two new cases to consider.

(iv) Suppose θ_i is $(t = t)$. In this case $\Gamma_{i+1} = \Gamma_i \cup \{(t = t)\}$. Assume for contradiction that there is a σ-derivation D of $\bot$ whose undischarged assumptions are in $\Gamma_i \cup \{(t = t)\}$. In D, put the right-hand label $(=I)$ on each leaf with left label $(t = t)$ and no dandah; in other words, turn all undischarged assumptions of this form into conclusions of $(=I)$. The result is a σ-derivation of $\bot$ from assumptions in Γ_i, which contradicts the inductive assumption that Γ_i is syntactically consistent.

(v) Suppose θ_i is $(\phi, x, s = t)$ where the sentences $(s = t)$ and $\phi[s/x]$ are in Γ_i. Then $\Gamma_{i+1} = \Gamma_i \cup \{\phi[t/x]\}$. Suppose for contradiction that there is a σ-derivation D with conclusion $\bot$ and undischarged assumptions all in $\Gamma_i \cup \{\phi[t/x]\}$. Every undischarged assumption of D of the form $\phi[t/x]$ can be derived from $(s = t)$ and $\phi[s/x]$ by $(=E)$, and this turns D into a σ-derivation of $\bot$ from assumptions in Γ_i. But by induction hypothesis there is no such derivation.

The claim is proved.

It remains to show that Δ is a Hintikka set. Properties (1)–(5) are satisfied as in the proof of Lemma 3.10.6.

For property (6), suppose t is a closed term of $\mathrm{LR}(\sigma)$. Then some θ_i is equal to $(t = t)$, and we chose Γ_{i+1} to be $\Gamma_i \cup \{(t = t)\}$. So $(t = t)$ is in Γ_{i+1}, hence in Δ, and property (6) holds.

For property (7), suppose ϕ is atomic and s, t are closed terms such that both $(s = t)$ and $\phi[s/x]$ are in Δ. Then $(s = t)$ is in Γ_i and $\phi[s/x]$ is in Γ_j for some i, j. Let k be the maximum of i, j, so that both sentences are in Γ_k. There is ℓ with $\ell \geq k$ such that θ_ℓ is $(\phi, x, s = t)$. We chose $\Gamma_{\ell+1}$ to be $\Gamma_\ell \cup \{\phi[t/x]\}$, so $\phi[t/x]$ is in $\Gamma_{\ell+1}$, hence in Δ. This shows (7) is satisfied and completes the proof that Δ is a Hintikka set. $\square$

As in Section 3.10, these two lemmas prove the Adequacy Theorem.

Theorem 5.10.5 (Completeness Theorem for qf Sentences) *Let Γ be a set of sentences of $LR(\sigma)$ and ψ a sentence of $LR(\sigma)$. Then*

$$\Gamma \vdash_\sigma \psi \quad \Leftrightarrow \quad \Gamma \models_\sigma \psi.$$

Proof This is the Soundness and Adequacy Theorems combined. □

Theorem 5.10.6 (Decidability Theorem for qf Sentences) *There is an algorithm which, given a finite set of qf sentences, will decide whether or not the set has a model.*

Proof We sketch the proof. Let ϕ be a qf sentence and ℓ its length. Let σ be the smallest signature so that ϕ is in $LR(\sigma)$; so σ is finite. By Exercise 5.6.5, if ϕ has a model then it has one with at most ℓ elements. Now we can list all the σ-structures with domain $\{1\}$, all the σ-structures with domain $\{1,2\}$, and so on up to those with domain $\{1,\dots,\ell\}$. The list is finite because σ is finite and there are only finitely many ways of interpreting each symbol in σ. We can check each listed σ-structure to see whether it is a model of ϕ. If none are, then ϕ has no model. □

When σ is a signature with no constant symbols, the Adequacy and Decidability Theorems still hold but for a trivial reason. The only qf sentences of $LR(\sigma)$ are formulas of propositional logic, built up from $\perp$. Such formulas are either tautologies or contradictions, so that the structures become irrelevant and the arguments of this section collapse down to truth table calculations.

Exercises

5.10.1. Justify (5.60) in full generality by establishing the following. Let Δ be a Hintikka set for $LR(\sigma)$, and define $\sim$ by (5.58). Let $s_1,\dots,s_n,t_1,\dots,t_n$ be closed terms such that $s_i \sim t_i$ whenever $1 \le i \le n$. Then:

    (a) If F is a function symbol of arity n, then $F(s_1,\dots,s_n) \sim F(t_1,\dots,t_n)$. [Show that $F(s_1,\dots,s_n) \sim F(t_1,\dots,t_{i-1},s_i,\dots,s_n)$ whenever $1 \le i \le n+1$, by induction on i. Each step uses (7) as in the proof of (5.62).]

    (b) If R is a relation symbol of arity n then $R(s_1,\dots,s_n) \in \Delta \Leftrightarrow R(t_1,\dots,t_n) \in \Delta$. [Use a similar induction argument.]

5.10.2. We write $N(\sigma,n)$ for the number of distinct σ-structures whose domain is the set $\{1,\dots,n\}$.

(a) Show that if σ consists of a single constant symbol then $N(\sigma, n) = n$.

(b) Show that if σ consists of a single k-ary relation symbol then $N(\sigma, n) = 2^{n^k}$.

(c) Show that if σ consists of a single k-ary function symbol then $N(\sigma, n) = n^{n^k}$.

(d) Show that if σ consists of the i distinct symbols $S_1, \ldots, S_i$ and we write $\{S_j\}$ for the signature with only the symbol S_j then $N(\sigma, n) = N(\{S_1\}, n) \times \cdots \times N(\{S_i\}, n)$.

6   Second interlude: the Linda problem

'Linda is 31 years old, single, outspoken, and very bright. She majored in philosophy. As a student, she was deeply concerned with issues of discrimination and social justice, and also participated in anti-nuclear demonstrations.

Please rank the following statements by their probability, using 1 for the most probable and 8 for the least probable.

(a) Linda is a teacher in an elementary school.

(b) Linda works in a bookstore and takes Yoga classes.

(c) Linda is active in the feminist movement.

(d) Linda is a psychiatric social worker.

(e) Linda is a member of the League of Women Voters.

(f) Linda is a bank teller.

(g) Linda is an insurance salesperson.

(h) Linda is a bank teller and is active in the feminist movement.'

Please write down your answers before reading on.

Logical analysis

In fact most of the questions were a blind. The only answers that interest us are those to the two questions

(f) Linda is a bank teller.
(h) Linda is a bank teller and is active in the feminist movement.

By ($\land$E) the statement

(h) Linda is a bank teller and is active in the feminist movement.

entails the statement

(f) Linda is a bank teller.

So (f) must be at least as probable as (h).

In our logic class the percentage of people who made (h) more probable than (f), even after completing Chapters 1–5, was 54%.

Psychological analysis

The quoted passage is from 'Judgments of and by representativeness' by Amos Tversky and Daniel Kahneman (pages 84–98 of Daniel Kahneman, Paul Slovic and Amos Tversky (eds), *Judgment under Uncertainty: Heuristics and Biases*, Cambridge University Press, Cambridge, 1982). It has given rise to a large number of psychological studies and analyses under the name of the *Linda Problem* or the *Conjunction Fallacy*. Tversky and Kahneman suggest that people tend to estimate (h) as more probable than (f) because (h) is more 'representative' of the kind of person that Linda is described as being.

Whatever the reason, this and several other examples given by Kahneman and Tversky seem to show that we often make choices incompatible with logic, even in situations where our choices have practical consequences. If most people making commercial decisions or playing the stock market commit systematic fallacies like this one, then anyone who can recognise and avoid the fallacies stands to make a killing. In 2002 Kahneman was awarded the Nobel Prize in Economics for his work with Tversky (who had died in 1996).

7　First-order logic

Now we turn to arguments that involve quantifier expressions 'There is' and 'For all', as we promised at the beginning of Chapter 5. We study what these expressions mean in mathematics (Sections 7.1–7.2), we build a formal semantics that gives necessary and sufficient conditions for a first-order sentence to be true in a structure (Section 7.3), we find natural deduction rules for these sentences (Section 7.4) and we prove a completeness theorem for the proof calculus got by adding these rules to those of earlier chapters (Section 7.6). The logic that we reach by adding these expressions and rules is called *first-order logic*. Sometimes you will also hear it called *predicate logic* or *elementary logic* or *classical logic*. Do not confuse it with *traditional logic*, which is the logic that was studied up till the middle of the nineteenth century—see (7.8) below.

Much of this chapter runs parallel to what we have already done in earlier chapters; but not all of it—there is a kind of mathematical lift-off when quantifiers become available. In Section 7.1 we will see how to express some arithmetic by using quantifiers and '='. Section 7.5 will take this idea much further, and show that a significant amount of arithmetic can be done by using first-order logic together with a small number of axioms expressing facts about the natural numbers. Many familiar classes of structures in algebra, such as groups, fields, rings or vector spaces, are defined by axioms that we can write in first-order logic. A set of first-order sentences used in order to define a class of structures is known as a *theory*, and in Section 7.7 we examine some theories.

One cannot go far in logic today without meeting infinity. In a first logic course you probably do not want to hear too much about infinity. So we postpone the details till Section 7.8, where they serve as a preparation for some brief remarks in Section 7.9 about the role of logic in the classification of mathematical structures.

7.1　Quantifiers

Definition 7.1.1 Suppose ϕ is a predicate and x a variable. Then we write

(7.1) $$\forall x \phi$$

for the statement 'For all x, ϕ'. For example,

(7.2) $$\forall x \ x \text{ is prime}$$

says 'For all elements x, x is prime'. This is mathematicians' talk; plain English says 'All elements are prime'.

We write

(7.3) $\exists x \phi$

for the statement 'For some x, ϕ', or equivalently 'There is x such that ϕ'. Thus

(7.4) $\exists x \ x$ is prime

is read 'There is an element x such that x is prime', or 'There is a prime element'.

In the terminology of Definitions 5.1.1 and 5.3.5, the universal quantifier $\forall x$ in (7.2) *binds* all the occurrences of x in (7.2) (including the occurrence of x in $\forall x$ itself). The same applies in (7.1). Likewise the existential quantifier $\exists x$ in (7.4) binds all occurrences of x in (7.4), and the same holds in (7.3). In Section 7.2 we will give a precise test of when an occurrence of a variable is bound.

Here are some examples of statements using quantifiers.

(7.5) $\exists y \ (y > x)$

says that some element is greater than x. Both occurrences of y in (7.5) are bound by the quantifier, but x is free because there is no quantifier $\forall x$ or $\exists x$.

(7.6) $\forall x \exists y \ (y > x)$

says that for every element x there is an element which is greater than x; in other words, there is no maximal element. All the occurrences of variables in (7.6) are bound.

(7.7) $\forall x \ (x$ is prime $\rightarrow x > 13)$

says that every element, if it is prime, is greater than 13; in other words, every prime element is greater than 13.

The next examples are from traditional logic; the letters A, E, I, O are medieval names for the sentence forms.

(7.8)

(A)	$\forall x \ (x$ is a man $\rightarrow x$ is mortal)	Every man is mortal.
(E)	$\forall x \ (x$ is a man $\rightarrow \neg \ (x$ is mortal))	No man is mortal.
(I)	$\exists x \ (x$ is a man $\wedge x$ is mortal)	Some man is mortal.
(O)	$\exists x \ (x$ is a man $\wedge \neg \ (x$ is mortal))	Some man is not mortal.

Note how $\forall$ goes with $\rightarrow$ and $\exists$ goes with $\wedge$ in these examples.

Algebraic laws are generally about 'all elements' of the relevant structure, so they use universal quantifiers:

(7.9) $\forall x \forall y \forall z \ (x + (y + z) = (x + y) + z)$

But a few laws say that some element exists, so they need an existential quantifier. For example, in a field, every nonzero element has a multiplicative inverse:

$$(7.10) \qquad \forall x \ (x \neq 0 \rightarrow \exists y \ (x \cdot y = 1))$$

In all these examples the quantifiers appear where we would expect to see them. But the next example is different: the English sentence that we are paraphrasing does not contain an 'all' or a 'some' or any similar phrase.

Bertrand Russell England, 1872–1970. Definite descriptions can be expressed with = and quantifiers.

We paraphrase

(7.11) $\qquad\qquad$ The snark is a boojum

thus, following Bertrand Russell:

(7.12)
- (1) There is at least one snark, and
- (2) there is at most one snark, and
- (3) every snark is a boojum.

Expressing this with quantifiers, we take the conjunction of three sentences:

(7.13)
- (1) $\exists x$ x is a snark.
- (2) $\forall x \forall y \ ((x$ is a snark $\wedge \ y$ is a snark$) \rightarrow x = y)$.
- (3) $\forall x \ (x$ is a snark $\rightarrow x$ is a boojum$)$.

Note how Russell's paraphrase uses the same predicate 'x is a snark' several times, and not always with the same variables.

In examples, such as (7.11)–(7.13), where the quantifiers do not simply translate phrases of the original statement, we say that the paraphrase (7.13) using quantifiers is an *analysis* of the original. Analyses of this kind are important at all levels of logic, even at the frontiers of research.

Exercises

7.1.1. Use quantifiers to express the following, using the predicates 'x is a triffid' ($T(x)$ for short) and 'x is a leg of y' ($L(x, y)$):

   (a)   There is at most one triffid. [See (7.12).]

   (b)   There are at least two triffids. [The negation of (a).]

   (c)   There are at least three triffids.

   (d)   There are exactly two triffids.

   (e)   Every triffid has at least two legs.

   (f)   Every triffid has exactly two legs.

   (g)   There are at least two triffids that have a leg. [Not necessarily the same leg!]

   (h)   There is a leg that at least two triffids have.

7.1.2. We write $\exists_{\geqslant n} x\, P(x)$ for 'There are at least n elements x such that $P(x)$'. For example, $\exists_{\geqslant 1} x\, P(x)$ says the same as $\exists x\, P(x)$. Show that, given any positive integer n, the statement $\exists_{\geqslant n+1} x P(x)$ can be expressed using $\exists$, $\exists_{\geqslant n}$ and $=$ (and of course $P(x)$).

7.1.3. Using quantifiers and the predicates 'x is male', 'x is female', 'x is a child of y', express the following:

   (a) x is a father.

   (b) x has a daughter.

   (c) x has exactly two children.

   (d) x is a grandparent of y.

   (e) x is a brother of y. (I.e. a full brother: both parents the same.)

   (f) x is a maternal aunt of y.

7.1.4. Using the predicate $(x < y)$ and quantifiers where needed, express the following:

   (a) x is less than or equal to y.

   (b)   If $x < y$ then there is an element strictly between x and y.

   (c) x is less than every other element.

   (d) x is a minimal element. (I.e. there is no element strictly less than x.)

    (e)   There is no least element.

    (f) y is the immediate successor of x. (In a linear order.)

    (g)   Every element has an immediate successor. (Again in a linear order.)

7.1.5.   In this exercise, F is a function from the set A to the set B, $A(x)$ means 'x is in the set A' and $B(x)$ means 'x is in the set B'. Express the following in mathematical English without quantifier symbols:

    (a) $\forall x \forall y \ (F(x) = F(y) \rightarrow x = y)$.

    (b) $\forall x \ (B(x) \rightarrow \exists y (F(y) = x))$.

    (c) $\forall x \ (B(x) \rightarrow \exists y \forall z \ ((F(z) = x) \leftrightarrow (z = y)))$.

    (d) $\exists x \ (F(x) = x)$.

    (e) $\forall x \ (A(x) \leftrightarrow B(x))$.

7.2 Scope and freedom

Definition 7.2.1 Let ψ be a formula of $\mathrm{LR}(\sigma)$.

(a) Consider an occurrence of a quantifier $\forall x$ or $\exists x$ in ψ. The *scope* of this occurrence is the subformula of ψ that starts at this occurrence.

(b) We say that an occurrence of a variable x in ψ is *bound in* ψ if it lies within the scope of a quantifier $\forall x$ or $\exists x$ with the same variable x. We say that the occurrence is *free in* ψ if it is not bound in ψ.

(c) A *sentence* of $\mathrm{LR}(\sigma)$ is a formula of $\mathrm{LR}(\sigma)$ in which no variable has a free occurrence. (This extends Definition 5.6.1 from qf formulas to the whole of LR.)

To see what Definition 7.2.1 is about, suppose this is the parsing tree of a formula ψ with a quantifier in it (with only the relevant details shown):

(7.14)

The scope of the $\forall x$ shown in the middle is the subformula $\forall x \phi$ associated with the node marked $\forall x$. Our explanation of the meaning of the quantifier $\forall x$ in Definition 7.1.1 was in terms of this subformula $\forall x \phi$. No occurrence of x in $\forall x \phi$ can be free in ψ. The same would apply if we had $\exists$ rather than $\forall$.

Note that in (7.14) there could be another quantifier $\forall x$, with the same variable x, inside ϕ. The occurrences of x in the scope of this second occurrence of $\forall x$ will still be bound in ψ, but not because of the first occurrence of $\forall x$ (since they are already bound in ϕ). The same would apply if there was an occurrence of $\exists x$ in ϕ.

Example 7.2.2 We determine which occurrences of variables are free and which are bound in the following formula (where c is a constant symbol, F is a function symbol and P is a relation symbol):

(7.15) $((x = y) \to \forall y(\exists z(x = F(z, c, w)) \wedge P(y, z)))$

Here is the parsing tree, taken as far as formulas that contain no quantifiers:

(7.16)

From the tree we can read off the scopes of the two quantifier occurrences; they are shown by the lines below the formula in (7.17). Then finally we can label the occurrences as F (free) or B (bound), as shown in (7.17). For example, the third occurrence of z is not in the scope of $\exists z$, so it is free.

$$
\begin{array}{llllll}
\text{F} \ \text{F} & \text{B} \ \ \text{B} \ \text{F} & \text{B} \ \ \text{F} & \text{B} \ \text{F}
\end{array}
$$

(7.17) $((x = y) \to \forall y(\exists z(x = F(z, c, w)) \wedge P(y, z)))$
                 ── y
                     ──────────────────────── z

After a little practice you should be able to work out, given any reasonable formula, which variable occurrences are free and which are bound.

Knowing which occurrences of variables are free and which are bound in (7.15), you can check what terms are substitutable for the variables in this formula (cf. Definition 5.2.9). For example, the term $F(y, c, c)$ *is not* substitutable for x in (7.15); this is because there is a free occurrence of x inside the scope of a quantifier with y. On the other hand, $F(z, c, c)$ *is* substitutable for y in (7.15), since there is no free occurrence of y inside the scope of a quantifier with z.

When a formula ϕ has free occurrences of variables, it does not make sense to ask whether ϕ is true in a given structure A, because we do not know what elements the variables stand for. But if ϕ is a sentence, then the explanations of quantifiers in Definition 7.1.1, together with our earlier explanations of logical symbols, completely determine whether ϕ makes a true statement about A when its symbols from σ are interpreted by means of A. Just as in Section 5.6, when ϕ does make a true statement about A, we say that A is a *model* of ϕ, or that ϕ is *true in* A, in symbols $\models_A \phi$. As after Definition 3.5.7, the process of checking whether a given sentence is true in a given structure is known as *model checking*.

Example 7.2.3 We do some model checking. Let ϕ be the following sentence of σ_{arith}:

$$\forall x_0 \exists x_1 \ (\cdot(x_0, x_1) = \bar{S}(\bar{0}))$$

or in shorthand

$$\forall x_0 \exists x_1 \ (x_0 x_1 = \bar{1}).$$

Is ϕ true in $\mathbb{N}/3\mathbb{N}$? We answer the question as follows.

Step One. The sentence begins with a quantifier $\forall x_0$. Removing this quantifier reveals the predicate

$$(7.18) \qquad \exists x_1 \ (x_0 \times x_1 = \bar{1})$$

which has one free occurrence of x_0. Since the domain of $\mathbb{N}/3\mathbb{N}$ consists of the three numbers $0, 1, 2$, Definition 7.1.1 tells us that ϕ is true in $\mathbb{N}/3\mathbb{N}$ if and only if all of the following three sentences are true in $\mathbb{N}/3\mathbb{N}$:

$$(7.19) \qquad \begin{array}{ll} (1) & \exists x_1 \ (\bar{0} \times x_1 = \bar{1}) \\ (2) & \exists x_1 \ (\bar{1} \times x_1 = \bar{1}) \\ (3) & \exists x_1 \ (\bar{2} \times x_1 = \bar{1}) \end{array}$$

Step Two. We check whether the sentences in (7.19) are true, starting with the first. Again we strip away the quantifier and find the predicate

$$(\bar{0} \times x_1 = \bar{1})$$

This predicate has a free occurrence of x_1. By Definition 7.1.1 again, the first sentence of (7.19) is true in $\mathbb{N}/3\mathbb{N}$ if and only if at least one of the following three sentences is true in $\mathbb{N}/3\mathbb{N}$:

$$(7.20) \qquad \begin{array}{ll} (1) & (\bar{0} \times \bar{0} = \bar{1}) \\ (2) & (\bar{0} \times \bar{1} = \bar{1}) \\ (3) & (\bar{0} \times \bar{2} = \bar{1}) \end{array}$$

Similarly for the other two sentences in (7.19).

Step Three. Now we check the truth of the atomic sentences in (7.20). As in Chapter 5, we evaluate in $\mathbb{N}/3\mathbb{N}$ the two sides of each equation, and we check whether they are equal. As it happens, each of the equations in (7.20) evaluates to 0 on the left and 1 on the right, so they are all false. Hence the first sentence of (7.19) is false, and this shows that the sentence ϕ is false in A. (If the first sentence of (7.19) had been true, we would have checked the second and if necessary the third.)

Remark 7.2.4 We could finish the model checking in Example 7.2.3 because the structure $\mathbb{N}/3\mathbb{N}$ has a finite domain. To do a similar calculation with $\mathbb{N}$ would involve checking infinitely many numbers, which is clearly impossible. There is no way around this problem. Gödel showed (Theorem 8.2) that there is no algorithm for determining whether a given sentence of $\mathrm{LR}(\sigma_{\mathrm{arith}})$ is true in $\mathbb{N}$.

Remark 7.2.5 Did you notice that the model checking in Example 7.2.3 went in the opposite direction from the way we assigned a truth value to a propositional formula in a structure? In the LP case we determined truth values by walking up the parsing tree, starting at the atomic formulas. But in Example 7.2.3 we started with the whole sentence and worked our way down to atomic sentences.

There are advantages to working from the outside inwards. The model checking procedure of Example 7.2.3 has been marketed as a game (*Tarski's world*, by Jon Barwise and John Etchemendy). The players are student and computer; together they strip down the sentence, and as they go they assign elements of the structure to the free variables. At the very least, the game is a good way of learning the symbolism of first-order logic.

Nevertheless, a proper mathematical definition of 'model' needs to go in the other direction, climbing up the parsing tree. We find such a definition in the next section and Appendix B. To prepare for it we now give a mathematical definition of $FV(\phi)$, the set of variables that have free occurrences in ϕ. The definition is recursive, and it starts by defining the set $FV(t)$ of variables that occur in a term t.

Definition 7.2.6 Let σ be a signature. We define $FV(t)$ when t is a term of $\mathrm{LR}(\sigma)$:

(a) If t is a constant symbol c then $FV(t) = \emptyset$.

(b) If t is a variable x then $FV(t) = \{x\}$.

(c) If t is $F(t_1, \ldots, t_n)$ where F is a function symbol of arity n and $t_1, \ldots, t_n$ are terms, then $FV(t) = FV(t_1) \cup \cdots \cup FV(t_n)$.

We define $FV(\phi)$ when ϕ is a formula of $\mathrm{LR}(\sigma)$:

(a) If ϕ is $R(t_1, \ldots, t_n)$ where R is a relation symbol of arity n and $t_1, \ldots, t_n$ are terms, then $FV(\phi) = FV(t_1) \cup \cdots \cup FV(t_n)$.

(b) If ϕ is $(s = t)$ where s, t are terms, then $FV(\phi) = FV(s) \cup FV(t)$.

(c) $FV(\bot) = \emptyset$.

(d) $FV((\neg\psi)) = FV(\psi)$.

(e) $FV((\psi \wedge \chi)) = FV((\psi \vee \chi)) = FV((\psi \rightarrow \chi)) = FV((\psi \leftrightarrow \chi)) = FV(\psi) \cup FV(\chi)$.

(f) $FV(\forall x\psi) = FV(\exists x\psi) = FV(\psi) \setminus \{x\}$. (Here $X \setminus Y$ means the set of all elements of X that are not elements of Y.)

Inspection shows that $FV(t)$ is the set of all variables that occur in t, and $FV(\phi)$ is the set of all variables that have free occurrences in ϕ. But note that the definition of FV never mentions occurrences or freedom. In the same spirit we can write a recursive definition of 'term t is substitutable for variable y in formula ϕ'; see Exercise 7.2.7.

We extend the definition of substitution (Definition 5.4.4) to first-order logic as follows.

Definition 7.2.7 Let E be a term or formula of LR. Let $y_1, \ldots, y_n$ be distinct variables and $t_1, \ldots, t_n$ terms such that each t_i is substitutable for y_i in E. Then $E[t_1/y_1, \ldots, t_n/y_n]$ is defined recursively as in Definition 5.4.4, but with the following new clause to cover formulas with quantifiers:

(e) Suppose E is $\forall x\phi$. If x is not in $\{y_1, \ldots, y_n\}$ then $E[t_1/y_1, \ldots, t_n/y_n]$ is

$$\forall x(\phi[t_1/y_1, \ldots, t_n/y_n])$$

(using extra parentheses to remove an ambiguity). If x is y_n then $E[t_1/y_1, \ldots, t_n/y_n]$ is

$$\forall x(\phi[t_1/y_1, \ldots, t_{n-1}/y_{n-1}])$$

and similarly for the other variables y_i. The same holds with $\exists x$ in place of $\forall x$.

Recall that $E[t_1/y_1, \ldots, t_n/y_n]$ is undefined unless each t_i is substitutable for y_i in E (Definition 5.2.10). Because of this, we need to check that in the recursive definition, if this condition is met on the left side of each clause then it is still met on the right side. For example, in (e) when x is y_n and $i < n$, the term t_i is substitutable for y_i in ϕ since it is substitutable for y_i in $\forall x\phi$, because the free occurrences of y_i in ϕ are exactly those in $\forall x\phi$ (since we assumed $y_i \neq y_n$), and all quantifier occurrences in ϕ are also in $\forall x\phi$.

Exercises

7.2.1. Identify the free and bound occurrences of variables in the following formulas of LR(σ). (Here c is a constant symbol, F, G are function symbols, and $P, Q, R, <$ are relation symbols. Some of the formulas are written in shorthand.)

(a) $\forall x(Q(F(F(y)), G(y, z)) \rightarrow \forall z R(x, y, F(F(F(z)))))$.

(b) $\exists x_1(\forall x_2 P(F(x_2)) \rightarrow Q(x_2, x_3))$.

(c) $((\exists x P(x, y) \vee \forall y P(y, z)) \rightarrow P(x, z))$.

(d) $x = c \wedge \forall x(P(G(G(x, y), y)) \rightarrow \exists y Q(y, G(x, x)))$.

(e) $(\forall x(x = y) \wedge \forall y(y = z)) \rightarrow (x = z)$.

(f) $\exists x \forall y(x < y) \vee \forall x(x = y)$.

7.2.2. The signature σ has 1-ary relation symbols A and P, a 2-ary relation symbol R and a constant symbol c. Which of the following formulas of LR(σ) are sentences?

(a) $\forall x(A(x) \wedge P(x)) \vee \neg(P(x))$.

(b) $A(c) \wedge (P(c) \rightarrow \forall x(x = c))$.

(c) $\forall z \ (\forall x R(x, z) \leftrightarrow \exists x R(x, x))$.

(d) $\forall y \exists z \ (R(y, c) \rightarrow R(x, z))$.

7.2.3. Use model checking to determine whether $\exists x(\bar{+}(\bar{\cdot}(x, x), \bar{S}(\bar{0})) = \bar{0})$ (or in shorthand $\exists x(x^2 + 1 = 0)$) is true (a) in $\mathbb{N}/5\mathbb{N}$, (b) in $\mathbb{N}/7\mathbb{N}$. [This exercise is asking you to determine whether -1 has a square root mod 5 or mod 7.]

7.2.4. (a) Adjust Definition 7.2.6 to give a definition of the set $BV(\phi)$ of all variables that have bound occurrences in ϕ.

(b) Use Definition 7.2.6 and your definition in (a) to *calculate* $FV(\phi)$ and $BV(\phi)$ for each of the following formulas ϕ:

(i) $\exists x_0(\forall x_1 P(F(x_1, x_2)) \wedge Q(x_0, x_1))$.

(ii) $\forall x \forall y P(x, y) \vee \forall x P(y, y)$.

(iii) $\forall x(R(x, y) \rightarrow \exists y W(x, y, w))$.

7.2.5. (a) Definition 7.2.6 of $FV(\phi)$ for formulas ϕ can be written as a compositional definition. Here is one of the compositional clauses:

where $\square \in \{\wedge, \vee, \rightarrow, \leftrightarrow\}$.

Write down the remaining clauses (assuming it has been done for terms).

(b) Do the same for your definition of BV in Exercise 7.2.4(a).

7.2.6. For each of the following formulas, determine whether the term $F(x, y)$ is substitutable for the variable x in the given formula.

(a) $((\forall y P(y, y) \rightarrow P(x, y)) \vee \exists y P(y, y))$.

(b) $(\forall y P(y, y) \rightarrow \exists y R(x, y))$.

(c) $\forall x (Q(x) \rightarrow \exists y P(y, x))$.

7.2.7. Let $\mathrm{Sub}(t, y, \phi)$ mean 'the term t is substitutable for the variable y in the formula ϕ'. Define $\mathrm{Sub}(t, y, \phi)$ by recursion on the complexity of ϕ. The first clause is

 • If ϕ is atomic then $\mathrm{Sub}(t, y, \phi)$.

(But the interesting clauses are those where ϕ starts with a quantifier.)

7.2.8. An important fact about first-order languages LR is that the scope of a quantifier occurrence never reaches outside the formula containing the occurrence. In natural languages this is not so; for example,

(7.21)   Every farmer has a donkey. He keeps an eye on it.

(The pronoun 'it', which functions as a variable, is bound by 'a donkey' in the previous sentence.) Because of examples like this (which go back to Walter Burley), this phenomenon is known as *donkey anaphora*. Find two examples of donkey anaphora in mathematical textbooks.

7.3 Semantics of first-order logic

Suppose σ is a signature, $\phi(y_1, \dots, y_n)$ is a formula of $\mathrm{LR}(\sigma)$, A a σ-structure and $(a_1, \dots, a_n)$ an n-tuple of elements of A. Recall from Section 5.7 that we say that $(a_1, \dots, a_n)$ *satisfies* ϕ *in* A, in symbols

(7.22) $\models_A \phi[\bar{a}_1 / y_1, \dots, \bar{a}_n / y_n]$

if the sentence $\phi[\bar{a}_1 / y_1, \dots, \bar{a}_n / y_n]$ is true in A, where each $\bar{a}_i$ is a constant symbol (if necessary a witness) naming a_i. We will carry this idea over from qf LR to first-order logic. There is a twist, first noticed by Alfred Tarski. We explained 'satisfies' in terms of 'true'. But in order to give a formal definition of *truth* for LR, it seems that we need to go in the opposite direction: first we define *satisfaction* by recursion on complexity, and then we come back to *truth* as a special case. Definition 7.3.1 will define satisfaction and then truth. Theorem 7.3.2 will confirm that after these notions have been set up, satisfaction can still be explained in terms of truth.

A temporary notational device will be useful. Suppose $(y_1, \dots, y_n)$ are distinct variables and x is a variable. We write $(y_1, \dots, y_n, x)$ for the sequence got from $(y_1, \dots, y_n)$ by adding x at the end, and *removing y_i if y_i was also the*

variable x. For example,

$$(\underline{x_2}, x_3, x_4) \text{ is } (x_2, x_3, x_4), \text{ but}$$
$$(\underline{x_2}, x_3, x_2) \text{ is } (x_3, x_2).$$

If $a_1, \ldots, a_n, b$ are elements correlated to $y_1, \ldots, y_n, x$, respectively, then we write $(\underline{a_1, \ldots, a_n}, b)$ for the sequence $(a_1, \ldots, a_n, b)$, but with a_i left out if x is y_i.

Definition 7.3.1 (Tarski's Truth Definition) Let σ be a signature, $\phi(y_1, \ldots, y_n)$ a formula of $\text{LR}(\sigma)$, A a σ-structure and $(a_1, \ldots, a_n)$ an n-tuple of elements of A. Then we define '$(a_1, \ldots, a_n)$ satisfies ϕ in A' by recursion on the complexity of ϕ. Clauses (a)–(f) of the definition are exactly as Definition 5.7.6. We add two more clauses for quantifiers.

(g) If ϕ is $\forall x\psi$, then $(a_1, \ldots, a_n)$ satisfies ϕ in A if and only if for every element a of A, $(\underline{a_1, \ldots, a_n}, a)$ satisfies $\psi(\underline{y_1, \ldots, y_n}, x)$.

(h) If ϕ is $\exists x\psi$, then $(a_1, \ldots, a_n)$ satisfies ϕ in A if and only if for some element a of A, $(\underline{a_1, \ldots, a_n}, a)$ satisfies $\psi(\underline{y_1, \ldots, y_n}, x)$.

Suppose ϕ is a sentence. Then we say ϕ is *true* in A, and that A is a *model* of ϕ, in symbols $\models_A \phi$, if the empty sequence () satisfies ϕ in A.

A special case of this definition is that if $\phi(x)$ is a formula of $\text{LR}(\sigma)$ and A is a σ-structure, then $\forall x\phi$ is true in A if and only if for every element a of A, the 1-tuple (a) satisfies ϕ. Ignoring the difference between a and (a), this agrees with our informal explanation of $\forall$ in Definition 7.1.1. The same goes for $\exists$.

Appendix B recasts Definition 7.3.1 as a compositional definition. In some sense, the question whether an n-tuple $(a_1, \ldots, a_n)$ satisfies a formula $\phi(y_1, \ldots, y_n)$ in a structure A can be calculated by tree-climbing, just like the truth values of complex formulas. But unfortunately the amount of information that has to be carried up the parsing tree is in general infinite, as we noted

Alfred Tarski Poland and USA, 1901–1983. We can give a mathematical definition of truth, though not for all languages at once.

in Remark 7.2.4; so literal calculation is out of the question except for finite structures.

Theorem 7.3.2 *Suppose A is a σ-structure and $\phi(y_1, \ldots, y_n)$ is a formula of $LR(\sigma)$. Let $a_1, \ldots, a_n$ be elements of A, and for each i let t_i be a closed term (possibly a witness) that names a_i in the sense that $(t_i)_A = a_i$. Then the following are equivalent:*

(a) $(a_1, \ldots, a_n)$ satisfies ϕ in A.

(b) $\phi[t_1/y_1, \ldots, t_n/y_n]$ is true in A.

Proof   This is proved by induction on the complexity of ϕ. Where ϕ is atomic, it holds by Definition 5.7.6(a). The truth function cases are straightforward. We give here the cases where ϕ begins with a quantifier, say ϕ is $\forall x \psi$ for some formula $\psi(y_1, \ldots, y_n, x)$. To avoid some fiddly details we assume x is not among $y_1, \ldots, y_n$, so that we can ignore the underlining. (Conscientious readers can fill in these details.) We write $\psi'(x)$ for $\psi[t_1/y_1, \ldots, t_n/y_n]$. Then the following are equivalent (where 'in A' is understood throughout):

(a) $(a_1, \ldots, a_n)$ satisfies $\forall x \psi$.
(b)   For every element a, $(a_1, \ldots, a_n, a)$ satisfies ψ.
(c)   For every closed term t, $\psi[t_1/y_1, \ldots, t_n/y_n, t/x]$ is true.
(d)   For every closed term t, $\psi'[t/x]$ is true.
(e)   For every element a, the 1-tuple (a) satisfies ψ'.
(f) $\forall x \psi'$ is true.
(g) $\phi[t_1/y_1, \ldots, t_n/y_n]$ is true.

Here (a) $\Leftrightarrow$ (b) and (e) $\Leftrightarrow$ (f) are by Definition 7.3.1; (b) $\Leftrightarrow$ (c) and (d) $\Leftrightarrow$ (e) are by induction hypothesis (since both ψ and ψ' have lower complexity than ϕ); (c) $\Leftrightarrow$ (d) is by the definition of ψ' and (f) $\Leftrightarrow$ (g) is by Definition 7.2.7.
The case with $\exists$ in place of $\forall$ is closely similar. $\square$

One consequence of Theorem 7.3.2 is that the truth of $\phi[t_1/y_1, \ldots, t_n/y_n]$ in A depends only on the elements $(t_1)_A, \ldots, (t_n)_A$, and not on the choice of closed terms $t_1, \ldots, t_n$ to name these elements.

The *Principle of Irrelevance* for LR says that the question whether or not $(a_1, \ldots, a_n)$ satisfies $\phi(y_1, \ldots, y_n)$ in A is not affected by (1) changing s_A where s is a symbol of the signature of A that does not occur in ϕ, or by (2) changing a_i when y_i does not have a free occurrence in ϕ. As far as (1) goes, it can be read off from the compositional definition of the semantics—but you can save yourself the effort of reading Appendix B by giving an argument by recursion on complexity. Matters are a little trickier with (2), and here we spell out just one case (though a representative one).

Lemma 7.3.3 *Let $\phi(y_1, \ldots, y_n)$ be a formula of $LR(\sigma)$ where $y_1, \ldots, y_n$ are the variables occurring free in ϕ, and let $\phi'(y_1, \ldots, y_{n+1})$ be the same formula* *but with one more variable listed. Let $a_1, \ldots, a_{n+1}$ be elements of A. Then $(a_1, \ldots, a_n)$ satisfies ϕ in A if and only if $(a_1, \ldots, a_{n+1})$ satisfies ϕ' in A.*

Proof The proof is by induction on the complexity of ϕ. The interesting case is where ϕ begins with a quantifier $\forall x$ or $\exists x$ which binds at least one other occurrence of x. Suppose ϕ is $\forall x \psi$ for some formula $\psi(y_1, \ldots, y_n, x)$. We note that x cannot be among the variables $y_1, \ldots, y_n$, but it could be y_{n+1}. To cover this possibility we write ψ as $\psi'(\underline{y_1, \ldots, y_{n+1}}, x)$ when y_{n+1} is considered as one of its listed variables.

Now $(a_1, \ldots, a_n)$ satisfies ϕ in A if and only if:

for every element $a, (a_1, \ldots, a_n, a)$ satisfies $\psi(y_1, \ldots, y_n, x)$.

By induction hypothesis if x is not y_{n+1}, and trivially if x is y_{n+1}, this is equivalent to

for every element $a, (\underline{a_1, \ldots, a_{n+1}}, a)$ satisfies $\psi'(\underline{y_1, \ldots, y_{n+1}}, x)$,

which in turn is equivalent to:

$(a_1, \ldots, a_{n+1})$ satisfies $\phi'(y_1, \ldots, y_{n+1})$.

A similar argument applies with $\exists$. □

Now that we have extended the notion of model from qf LR to LR, several other notions generalise at once. First-order sentences are gregarious animals; they like to travel in sets. We will see a number of examples of this in Section 7.7. So it makes sense to generalise Definition 5.6.7 not just from qf LR to LR, but also from single sentences to sets of sentences.

Definition 7.3.4 Let σ be a signature.

(a) A set of sentences of LR(σ) is called a *theory*.

(b) If Γ is a theory in LR(σ), then we say that a σ-structure A is a *model* of Γ if A is a model of every sentence in Γ. Given a sentence ψ of LR(σ), we write

(7.23) $\Gamma \models_\sigma \psi$

to mean that every σ-structure that is a model of Γ is also a model of ψ. We call (7.23) a *semantic sequent*. A special case is where Γ is empty; we say that ψ is *valid*, in symbols $\models_\sigma \psi$, if every σ-structure is a model of ψ.

(c) By a *counterexample* to the sequent $(\Gamma \vdash_\sigma \psi)$ we mean a σ-structure that is a model of Γ but not of ψ. (So one way to show that a semantic sequent is not correct is to construct a counterexample to it.)

(d) We say that a theory Γ is *consistent* or *satisfiable* if at least one σ-structure is a model of Γ. We say that Γ is *inconsistent* if no σ-structure is a model of Γ.

Exercise 5.6.4 can be proved for the whole of LR, not just for qf LR. In this sense, the subscript $_\sigma$ in $\models_\sigma$ in the definition above is redundant, and we will usually leave it out unless we want to emphasise that we are discussing σ-structures.

Theorem 7.3.5 *Let σ be a signature and let $\phi(y_1,\ldots,y_n)$ and $\psi(y_1,\ldots,y_n)$ be formulas of $LR(\sigma)$. Suppose A is a σ-structure. Then the following are equivalent:*

(i) If $(a_1,\ldots,a_n)$ is any n-tuple of elements of A, then $(a_1,\ldots,a_n)$ satisfies ϕ in A if and only if it satisfies ψ in A.

(ii) $\models_A \forall y_1 \cdots \forall y_n\ (\phi \leftrightarrow \psi)$.

Proof By the clause for $\leftrightarrow$ in Definition 7.3.1 (in fact Definition 5.7.6(d)), (i) is equivalent to the statement

$$(7.24) \qquad \text{Every } n\text{-tuple } (a_1,\ldots,a_n) \text{ satisfies } (\phi \leftrightarrow \psi) \text{ in } A$$

This in turn is equivalent to the statement that for every $(n-1)$-tuple $(a_1,\ldots,a_{n-1})$,

$$(7.25) \qquad \text{For every element } a_n \text{ of } A, (a_1,\ldots,a_n) \text{ satisfies } (\phi \leftrightarrow \psi) \text{ in } A$$

and (7.25) is equivalent by Definition 7.3.1(g) to

$$(7.26) \qquad \text{Every } (n-1)\text{-tuple } (a_1,\ldots a_{n-1}) \text{ satisfies } \forall y_n(\phi \leftrightarrow \psi) \text{ in } A$$

Applying Definition 7.3.1(g) n times, we reach (ii). $\square$

Definition 7.3.6 When the conditions (i), (ii) of Theorem 7.3.5 hold, we say that ϕ and ψ are *equivalent in A*. We say that ϕ and ψ are *logically equivalent* (in symbols ϕ eq ψ) if they are equivalent in all σ-structures. Equivalence in A and logical equivalence are both equivalence relations on the set of formulas of $LR(\sigma)$.

Note that when we say two formulas ϕ and ψ are logically equivalent, we always intend $\phi(y_1,\ldots,y_n)$ and $\psi(y_1,\ldots,y_n)$ with the same listed variables. By the Principle of Irrelevance, any choice of variables will do provided it contains all variables occurring free in either ϕ or ψ.

We list some useful logical equivalences:

Theorem 7.3.7

(a) $\exists x \phi$ eq $\neg \forall x \neg \phi$.

(b) $\forall x \phi$ eq $\neg \exists x \neg \phi$.

(c) $\forall x(\phi \wedge \psi)$ *eq* $(\forall x\phi \wedge \forall x\psi)$.

(d) $\exists x(\phi \vee \psi)$ *eq* $(\exists x\phi \vee \exists x\psi)$.

(e) $\forall x\forall y\phi$ *eq* $\forall y\forall x\phi$.

(f) $\exists x\exists y\phi$ *eq* $\exists y\exists x\phi$.

(g) If the variable x has no free occurrences in ϕ then all of ϕ, $\forall x\phi$ and $\exists x\phi$ are logically equivalent.

(h) Suppose the variable y has no free occurrences in ϕ, and is substitutable for x in ϕ. Then $\forall x\phi$ eq $\forall y\phi[y/x]$, and $\exists x\phi$ eq $\exists y\phi[y/x]$.

(i) If the variable x has no free occurrences in ϕ then $\forall x(\phi \wedge \psi)$ eq $(\phi \wedge \forall x\psi)$ and $\forall x(\psi \wedge \phi)$ eq $(\forall x\psi \wedge \phi)$. This equivalence still holds if we make either or both of the following changes: (1) Put $\exists$ in place of $\forall$; (2) put $\vee$ in place of $\wedge$. (WARNING: The theorem does not claim that the equivalences hold with $\rightarrow$ or $\leftarrow$ in place of $\wedge$.)

Proof   As a sample we take (d) and (h).

(d) Let $\phi(y_1, \ldots, y_n, x)$ and $\psi(y_1, \ldots, y_n, x)$ be formulas of $\text{LR}(\sigma)$, where $y_1, \ldots, y_n$ lists the variables with free occurrences in $\exists x(\phi \vee \psi)$. Let A be a σ-structure and $(a_1, \ldots, a_n)$ an n-tuple of elements of A. Then the following are equivalent:

(i) $(a_1, \ldots, a_n)$ satisfies $(\exists x(\phi \vee \psi))(y_1, \ldots, y_n)$ in A.

(ii) For some element a of A, $(a_1, \ldots, a_n, a)$ satisfies $(\phi \vee \psi)(y_1, \ldots, y_n, x)$ in A.

(iii) For some element a of A, either $(a_1, \ldots, a_n, a)$ satisfies ϕ in A or it satisfies ψ in A.

(iv) Either for some element a of A, $(a_1, \ldots, a_n, a)$ satisfies ϕ in A, or for some element a of A, $(a_1, \ldots, a_n, a)$ satisfies ψ in A.

(v)   Either $(a_1, \ldots, a_n)$ satisfies $(\exists x\phi)(y_1, \ldots, y_n)$ in A or it satisfies $(\exists x\psi)(y_1, \ldots, y_n)$ in A.

(vi) $(a_1, \ldots, a_n)$ satisfies $(\exists x\phi \vee \exists x\psi)(y_1, \ldots, y_n)$ in A.

(h) For simplicity, we suppose that x is the only free variable in $\phi(x)$. Then y is the only free variable in $\phi[y/x]$, and we can write $\phi[y/x]$ as $\psi(y)$. The following are equivalent:

(i) $\forall x\phi$ is true in A.

(ii) For every closed term t naming an element of A, $\phi[t/x]$ is true in A.

(iii) For every closed term t naming an element of A, $\psi[t/y]$ is true in A.

(iv) $\forall y\psi$ is true in A.

The equivalence (ii) $\Leftrightarrow$ (iii) holds because the conditions on y imply that $\phi[t/x]$ and $\psi[t/y]$ are one and the same formula. $\square$

The Replacement Theorem for Propositional Logic (Theorem 3.7.6(b)) generalises to first-order logic. It tells us that if ϕ is logically equivalent to ψ, and a formula χ' comes from a formula χ by replacing a subformula of the form ϕ by one of the form ψ, then χ and χ' are logically equivalent. So, for example, we can use the logical equivalences in Exercise 3.6.2 to find, for every first-order formula χ, a formula χ' that is logically equivalent but does not use certain truth function symbols, along the lines of Example 3.7.8. In particular, each first-order formula is logically equivalent to one not containing either $\rightarrow$ or $\leftrightarrow$. But with Theorem 7.3.7 some new kinds of logical equivalence become available, as illustrated in the next two examples. (WARNING: If some variable occurs free in ψ but not in ϕ, it can happen that χ is a sentence but χ' is not.)

Example 7.3.8 (Changing bound variables) Take the formula

$$(7.27) \qquad \forall y(y < x \rightarrow \exists x R(x, y))$$

The first occurrence of x here is free but the second and third are bound. This is ugly; it goes against our feeling that the same variable should not be used to talk about different things in the same context. We can repair it as follows. Take a variable z that occurs nowhere in (7.27). By Theorem 7.3.7(h), $\exists x R(x, y)$ eq $\exists z R(z, y)$. So by the Replacement Theorem, (7.27) is logically equivalent to $\forall y(y < x \rightarrow \exists z R(z, y))$.

Example 7.3.9 (Prenex formulas) Take the formula

$$(7.28) \qquad (\exists x R(x) \wedge \exists x \neg R(x))$$

This formula is logically equivalent to each of the following in turn:

$\qquad (\exists x R(x) \wedge \exists z \neg R(z))$ as in Example 7.3.8
$\qquad \exists z (\exists x R(x) \wedge \neg R(z))$ by Theorem 7.3.7(i)
$\qquad \exists z \exists x (R(x) \wedge \neg R(z))$ by applying Theorem 7.3.7(i) to
$\qquad\qquad\qquad\qquad\qquad\qquad (\exists x R(x) \wedge \neg R(z))$,
$\qquad\qquad\qquad\qquad\qquad\qquad$ then using Replacement Theorem

A formula is said to be *prenex* if either it is quantifier-free, or it comes from a quantifier-free (qf) formula by adding quantifiers at the beginning. So we have shown that (7.28) is logically equivalent to the prenex formula

$$(7.29) \qquad \exists z \exists x (R(x) \wedge \neg R(z))$$

In fact, every formula of LR is logically equivalent to a prenex formula, by applying Theorem 7.3.7 repeatedly as above. For example, the formula $\neg\forall x\phi$ is logically equivalent to $\neg\forall x\neg\neg\phi$ [using the Double Negation Law (Example 3.6.5) and Replacement Theorem] and hence to $\exists x\neg\phi$ (by Theorem 7.3.7(a)). Likewise, $(\forall xR(x) \to Q(w))$ eq $((\neg\forall xR(x)) \vee Q(w))$ (Exercise 3.6.2(c)), eq $(\exists x\neg R(x) \vee Q(w))$ (as above and using Replacement), eq $\exists x(\neg R(x) \vee Q(w))$ (by Theorem 7.3.7(i)).

Exercises

7.3.1. Show that none of the following semantic sequents are correct, by constructing counterexamples to them.

(a) $\{P(\bar{0}), \forall x(P(x) \to P(S(x)))\} \models \forall xP(x)$.

(b) $\{\forall x\exists yR(x,y)\} \models \exists y\forall xR(x,y)$.

(c) $\{\forall x\exists y(F(y) = x)\} \models \forall x\forall y(F(x) = F(y) \to x = y)$.

(d) $\{\forall x(P(x) \to Q(x)), \forall x(P(x) \to R(x))\} \models \exists x(Q(x) \wedge R(x))$.

[See section 5.5 on how to describe a structure.]

7.3.2. Which of the following formulas are logically equivalent to which? Give reasons.

(a) $x = y$.

(b) $x = y \wedge w = w$.

(c) $x = y \wedge R(x)$.

(d) $x = x \wedge w = w$.

(e) $(x = y \wedge R(x)) \vee (y = x \wedge \neg R(x))$.

(f) $(x = y \wedge \forall x(x = x))$.

7.3.3. Find a formula in which no variable occurs both free and bound, which is logically equivalent to the following formula.

$$P(x,y) \vee \forall x\exists y \ (R(y,x) \to \exists xR(x,y))$$

7.3.4. For each of the following formulas, find a logically equivalent prenex formula.

(a) $R(x) \to (R(y) \to \exists xP(x,y))$.

(b) $\exists xP(x) \to \exists xQ(x)$.

(c) $\exists xP(x) \leftrightarrow \exists xQ(x)$.

(d) $\exists x\forall yR(x,y) \vee \forall x\exists y \ Q(x,y)$.

7.3.5. Prove (a)–(c), (e)–(g) and (i) of Theorem 7.3.7.

7.3.6. Show that if t is a term, x a variable and ϕ a formula, then there is a formula ψ logically equivalent to ϕ, such that t is substitutable for x in ψ. [Use the method of Example 7.3.8.]

7.3.7. Show: If ϕ eq ψ, x is a variable and c is a constant symbol, then $\phi[c/x]$ eq $\psi[c/x]$.

7.4 Natural deduction for first-order logic

Natural deduction for first-order logic works the same way as for LP and LR, except that we have new rules for introducing and eliminating the quantifiers.

We will need four new rules: (∀I), (∀E), (∃I) and (∃E). In these rules the distinction between free and bound occurrences of variables becomes important. When a formula has free variables, we treat them (for purposes of proving things) as if they were constant symbols.

The first two rules hardly need any mathematical examples to illustrate them. The first rests on the fact that if something is true for all elements, then it is true for any named element. The second rests on the fact that if something is true for a particular named element, then it is true for at least one element.

NATURAL DEDUCTION RULE (∀E)   Given a derivation

$$D$$
$$\forall x \phi$$

and a term t that is substitutable for x in ϕ, the following is also a derivation:

$$D$$
$$\frac{\forall x \phi}{\phi[t/x]}$$

Its undischarged assumptions are those of D.

NATURAL DEDUCTION RULE (∃I)   Given a derivation

$$D$$
$$\phi[t/x]$$

where t is a term that is substitutable for x in ϕ, the following is also a derivation·

$$D$$
$$\frac{\phi[t/x]}{\exists x \phi}$$

Its undischarged assumptions are those of D.

Example 7.4.1 Since $(x = x)$ is also $(x = x)[x/x]$, we have the derivation

$$
\text{(7.30)} \qquad\qquad
\cfrac{\cfrac{\forall x(x = x)}{(x = x)}\ (\forall\text{E})}{\exists x(x = x)}\ (\exists\text{I})
$$

showing that $\{\forall x(x = x)\} \vdash \exists x(x = x)$. Now everything is always identical with itself, so every structure is a model of $\forall x(x = x)$. But if the domain of A was allowed to be the empty set, then $\exists x(x = x)$ would be false in A. So we would have derived a false statement from a true one, and the natural deduction calculus would have failed the Soundness property. This is one reason why in Definition 5.5.2 we required the domain of a structure to be a non-empty set.

Not everybody is happy that the natural deduction calculus manages to prove the existence of something from nothing. There are proof calculi (e.g. forms of the tableau calculus) that avoid this. Sometimes in mathematics one would like to allow structures with empty domains, but these occasions are too few to justify abandoning natural deduction.

We turn to the introduction rule for $\forall$. Suppose ϕ is a predicate with a free variable x. To prove that ϕ is true for every element x, a standard move is to write

(7.31)   Let c be an element.

and then prove $\phi[c/x]$ *without using any information about c*. Then $\phi[c/x]$ must be true regardless of what element c is, so $\forall x\phi$ is proved.

For the natural deduction form of this move, we require that c does not occur in ϕ or in any of the assumptions used to deduce $\phi[c/x]$. This is a formal way of ensuring that we have not used any information about c in deriving $\phi[c/x]$. There is no need to discharge the assumption (7.31), because we will never need to state this assumption in a derivation.

In (7.31) we could have equally well used a variable y in place of c. For our formal rule we make the same condition on the variable y as we did on the constant c, namely that it never occurs in ϕ or in the assumptions from which $\phi[y/x]$ is derived.

NATURAL DEDUCTION RULE ($\forall$I)   If

$$
D
$$
$$
\phi[t/x]
$$

is a derivation, t is a constant symbol or a variable, and t does not occur in ϕ or in any undischarged assumption of D, then

$$\frac{\begin{array}{c} D \\ \phi[t/x] \end{array}}{\forall x \phi}$$

is also a derivation. Its undischarged assumptions are those of D.

We turn to ($\exists$E). This is the most complicated of the natural deduction rules, and many first courses in logic omit it. We will never use it after this section.

How can we deduce something from the assumption that 'There is a snark'? If we are mathematicians, we start by writing

(7.32)   Let c be a snark.

and then we use the assumption that c is a snark in order to derive further statements, for example,

(7.33) c is a boojum.

If we could be sure that (7.33) rests only on the assumption that there is a snark, and not on the stronger assumption that there is a snark called 'c', then we could discharge the assumption (7.32) and derive (7.33). Unfortunately, (7.33) does explicitly mention c, so we cannot rule out that it depends on the stronger assumption. Even if it did not mention c, there are two other things we should take care of. First, none of the other assumptions should mention c; otherwise (7.33) would tacitly be saying that some other element already mentioned is a snark, and this was no part of the assumption that there is a snark. Second, the name c should be free of any implications; for example, we cannot write $\cos \theta$ in place of c, because then we would be assuming that some value of $\cos$ is a snark, and again this assumes more than that there is a snark. However, these are all the points that we need to take on board in a rule for eliminating $\exists$. As with ($\forall$I), we allow a variable in place of the constant c, with the same conditions applying.

NATURAL DEDUCTION RULE ($\exists$E) If

$$\begin{array}{c} D \\ \exists x \phi \end{array} \quad \text{and} \quad \begin{array}{c} D' \\ \chi \end{array}$$

are derivations and t is a constant symbol or a variable which does not occur in χ, ϕ or any undischarged assumption of D' except $\phi[t/x]$, then

$$
\begin{array}{cc}
 & \phi[t/x] \\
D & D' \\
\exists x\phi & \chi \\
\hline
 & \chi
\end{array}
$$

is also a derivation. Its undischarged assumptions are those of D' except possibly $\phi[t/x]$, together with those of D.

Example 7.4.2 We prove the sequent

$$\{\forall x(P(x) \to Q(x)), \forall x(Q(x) \to R(x))\} \vdash \forall x(P(x) \to R(x))$$

$$
\cfrac{
\cfrac{
P(c)^{①} \quad \cfrac{\cfrac{\forall x(P(x) \to Q(x))}{(P(c) \to Q(c))}\ (\forall\mathrm{E})}{Q(c)}\ (\to\mathrm{E})
\qquad
\cfrac{\cfrac{\forall x(Q(x) \to R(x))}{(Q(c) \to R(c))}\ (\forall\mathrm{E})}{}
}{
\cfrac{R(c)}{(P(c) \to R(c))}\ (\to\mathrm{I})\ ①
}\ (\to\mathrm{E})
}{\forall x(P(x) \to R(x))}\ (\forall\mathrm{I})
$$

Note how we used a witness c in order to apply $(\forall\mathrm{I})$ at the final step.

Example 7.4.3 We prove the sequent $\{\exists x \forall y R(x,y)\} \vdash \forall y \exists x R(x,y)$.

(7.34)

$$
①\ \cfrac{
\exists x \forall y R(x,y) \qquad
\cfrac{
\cfrac{
\cfrac{
\cfrac{\forall y R(u,y)^{①}}{R(u,z)}\ (\forall\mathrm{E})
}{\exists x R(x,z)}\ (\exists\mathrm{I})
}{\forall y \exists x R(x,y)}\ (\forall\mathrm{I})
}{}
}{\forall y \exists x R(x,y)}\ (\exists\mathrm{E})
$$

We should check the conditions on $(\forall\mathrm{I})$ and $(\exists\mathrm{E})$. It is important to remember that the undischarged assumptions referred to in these rules are not the undischarged assumptions of (7.34); *they are the assumptions that are not yet discharged at a certain node in the derivation.* So the condition we have to check for the application of $(\forall\mathrm{I})$ is that z does not occur in $\exists x R(x,y)$ or in $\forall y R(u,y)$;

the fact that $\forall y R(u, y)$ is discharged in (7.34) is irrelevant, because it was discharged at the application of ($\exists$E), which comes below ($\forall$I). At the application of ($\exists$E) we have to check that u does not occur in $\forall y \exists x R(x, y)$ or in $\forall y R(x, y)$; the formula $\forall y R(u, y)$ is undischarged at the occurrence of $\forall y \exists x R(x, y)$ in the right-hand column, but the conditions on ($\exists$E) allow $\forall y R(u, y)$ to contain u.

At this point one expects a definition of σ-derivations, telling us how to recognise whether a diagram is or is not a σ-derivation. But Example 7.4.3 shows that we have a problem here. When we check whether rule ($\forall$I) or rule ($\exists$E) has been applied correctly, we need to know which are the undischarged assumptions *at the place where the rule is applied*, not the undischarged assumptions of the whole derivation. So we cannot follow the strategy of Definitions 3.4.1 and 5.4.5, first checking the applications of the rules and then identifying the undischarged assumptions. A subtler approach is needed.

The simplest remedy, and the usual one, is to recast the definition of derivations so that the dandah numbers become parts of the derivation. Each node will have three labels. As before, the left label at a node is a formula and the right node states the rule that justifies the formula. The third label—maybe a front label—indicates whether the formula on the left carries a dandah, and if so gives the number on the dandah; it also indicates whether the rule on the right is used to put dandahs on other formulas, and if so it gives the number on these dandahs. Now when we check a rule at a node ν, we use the front labels on higher nodes as raw data, to tell us which are the undischarged assumptions at ν. The correctness of these front labels is checked elsewhere, when we check the rules where the formulas were discharged. So everything gets checked, and there is no circularity.

It should be clear in principle how to write down a definition of 'σ-derivation' along these lines. But this is as far as we will go towards a definition in this book. If you have some experience in programming, you can probably see how to code up a tree with three labels on each node. Then you can write an exact definition of σ-derivation in the form of a computer program to check whether a given tree is or is not a σ-derivation. As a computing project this is not trivial but not particularly hard either. At the end of the project you can claim to have a precise and thorough proof of the next theorem.

Theorem 7.4.4 *Let σ be a signature that is a computable set of symbols (e.g. a finite set). There is an algorithm that, given any diagram, will determine in a finite amount of time whether or not the diagram is a σ-derivation.*

Proof Exactly as the proof of Theorem 3.4.2, but using a definition of σ-derivations as three-labelled trees satisfying appropriate conditions matching the natural deduction rules. $\square$

Definition 7.4.5 The definitions of σ-*sequent*, *conclusion* of a sequent, *assumptions* of a sequent, *correctness* of a sequent, a derivation *proving* a sequent, and *syntactically consistent* set of sentences are all as in Definition 5.4.7, but using a suitable update of Definition 5.4.5 for LR.

When Γ is a set of qf sentences and ψ is a qf sentence, we have one definition of $\Gamma \vdash_\sigma \psi$ in Definition 5.4.7 and another in Definition 7.4.5. The two definitions are not the same; in the sense of Definition 7.4.5, the sequent $(\Gamma \vdash_\sigma \psi)$ could be proved by a derivation using formulas that are not quantifier-free. Fortunately, this is only a temporary problem. The Soundness Theorem for LR (Theorem 7.6.1) will show that if there is such a derivation then $\Gamma \models \psi$; so it follows from the Adequacy Theorem for qf LR (Theorem 5.10.1) that the sequent $(\Gamma \vdash_\sigma \psi)$ is proved by some derivation containing only qf sentences.

Our final result in this section shows that derivations can be converted into other derivations that prove the same sequent, but using only formulas of some restricted kind.

Theorem 7.4.6

(a) *Suppose $\rho \subseteq \sigma$, and all the symbols in σ but not in ρ are constant symbols. Let Γ be a set of formulas of $LR(\rho)$ and ψ a formula of $LR(\rho)$. Then $\Gamma \vdash_\rho \psi$ if and only if $\Gamma \vdash_\sigma \psi$.*

(b) *Suppose σ is a signature with infinitely many constant symbols. Let Γ be a theory in $LR(\sigma)$ and ψ a sentence of $LR(\sigma)$, such that $\Gamma \vdash_\sigma \psi$. Then there is a σ-derivation D proving $(\Gamma \vdash_\sigma \psi)$, such that all formulas in D are sentences.*

Proof   (a) From left to right is straightforward: if D is a ρ-derivation then D is also a σ-derivation, from the definition of σ-derivations. For the converse, we suppose we have a σ-derivation D with conclusion ψ and whose undischarged assumptions are all in Γ. We will construct a ρ-derivation D' with the same conclusion and undischarged assumptions as D. Since D is a labelled planar graph, it uses only finitely many symbols. Hence D uses at most finitely many symbols that are in σ but not in ρ. By assumption these symbols are constant symbols; we list them without repetition as $c_1, \ldots, c_n$. Again because D uses only finitely many symbols, the fact that LR has infinitely many variables (Definition 5.3.3(a)) guarantees that there are distinct variables $y_1, \ldots, y_n$, none of which are used in D.

We construct a diagram D_1 as follows. For each term t, we write t' for the term got from t by replacing each occurrence of c_1 by an occurrence of y_1; likewise we write θ' for the result of the same replacement in a formula θ. (One can define t' and θ' recursively from t and θ and prove that t' is a term and θ' is a formula. The recursive definition is a slight simplification of that in the solution to Exercise 5.4.8, with a clause added for quantifiers.) The diagram D_1 is D with each left label θ replaced by θ'. We can check that D_1 is a σ-derivation. This means checking each node to see that the rule named in the right-hand label on the node is correctly applied. For the propositional rules this is fairly trivial, since the replacement of c_1 by y_1 leaves the truth function symbols unchanged. Also (=I) is very straightforward. We will examine (=E) and the rules for $\forall$.

First, let ν be a node of D carrying the right label (=E). Then at node ν in D' we have the formula $\phi[t/x]$, and at its daughters we have $\phi[s/x]$ and $(s = t)$, where s and t are substitutable for x in ϕ. The variable x serves only to mark the place where substitutions are made, so we can choose it to be a variable that never appears in any formulas of D or D', or in s or t. By construction the corresponding nodes in D' have the formulas $\phi[t/x]'$, $\phi[s/x]'$ and $(s = t)'$. We need to show that $(s = t)'$ is $(s' = t')$, $\phi[t/x]'$ is $\phi'[t'/x]$ and $\phi[s/x]'$ is $\phi'[s'/x]$. By the definition of the function $\theta \mapsto \theta'$, $(s = t)'$ is $(s' = t')$. The cases of $\phi[t/x]'$ and $\phi[s/x]'$ are similar, so we need prove only one. The first thing to check is that t' is substitutable for x in ϕ'; otherwise $\phi'[t'/x]$ is meaningless. If t' is not substitutable for x in ϕ, then some quantifier occurrence $\forall y$ or $\exists y$ in ϕ' has within its scope a free occurrence of x, and the variable y occurs in t'. Since the variable x never appears in any formula of D, there are no quantifiers $\forall x$ or $\exists x$ in ϕ, so x has a free occurrence in ϕ. Hence t does occur in $\phi[t/x]$, and so y_1 is different from any variable in t, and in particular it is not y. By the definitions of ϕ' and t' it follows that y already occurred in t and that x already occurred free inside the scope of $\forall y$ or $\exists y$ in ϕ; so t was not substitutable for x in ϕ, contrary to assumption. This establishes that $\phi'[t'/x]$ makes sense. The proof that it is equal to $\phi[t/x]'$ proceeds like the corresponding argument in Exercise 5.4.7.

Second, let ν be a node of D carrying the right label ($\forall$E). Then the unique daughter of ν carries a formula $\forall x\phi$ and ν itself has the formula $\phi[t/x]$, where t is some term substitutable for x in ϕ. In D' the corresponding formulas are $(\forall x\phi)'$ and $\phi[t/x]'$. We have to show that $(\forall x\phi)'$ is $\forall x(\phi')$, that t' is substitutable for x in ϕ' and that $\phi[t/x]' = \phi'[t'/x]$. The fact that $(\forall x\phi)'$ is $\forall x(\phi')$ is immediate from the definition of $\theta \mapsto \theta'$. For the rest, there are two cases according to whether or not x does have a free occurrence in ϕ. If it does not, then $\phi[t/x]$ is ϕ; also $\phi'[t'/x]$ is ϕ', since y_1 was chosen different from every variable occurring in D, so that ϕ' does not contain any free occurrence of x either. Then $\phi[t/x]' = \phi' = \phi'[t'/x]$ as required. If x does have a free occurrence in ϕ then t appears in $\phi[t/x]$, so we know that y_1 is distinct from every variable in t. The rest of this case is like the previous one.

Third, let ν be a node of D carrying the right label ($\forall$I). Then there are a formula ϕ, a term t and a variable z such that ν has left label $\forall z\phi$, the daughter ν' of ν has left label $\phi[t/z]$ and t is a constant symbol or variable which occurs nowhere in ϕ or any formula that is undischarged at ν'. If t is not c_1 then this condition is still met when each formula θ is replaced by θ'. If t is c_1, then the condition still holds but with y_1 in place of c_1. The other matters to check are similar to the previous case.

In sum, D_1 is a σ-derivation. Since c_1 never appears in ψ or any formula in Γ, D_1 is again a derivation proving $(\Gamma \vdash_\sigma \psi)$. By construction the only symbols of σ that are in D_1 but not in ρ are $c_2, \dots, c_n$. So by repeating the construction n times to construct σ-derivations $D_1, \dots, D_n$, we can remove all the constant symbols $c_1, \dots, c_n$. Then D_n will be a ρ-derivation proving $(\Gamma \vdash_\rho \psi)$.

(b) is proved similarly. Given a σ-derivation D that proves $(\Gamma \vdash_\sigma \psi)$, we list as $y_1, \dots, y_n$ the variables that have free occurrences in formulas of D, and we find distinct constant symbols $c_1, \dots, c_n$ of σ that never appear in D. Then we replace each formula θ in D by $\theta[c_1/y_1]$, etc. The argument is essentially the same as that for (a), because the conditions on the term t in $(\forall I)$ and $(\exists E)$ are exactly the same regardless of whether t is a variable or a constant symbol. $\square$

Remark 7.4.7 The two rules $(\forall I)$ and $(\exists E)$ are often given in a more general form with weaker conditions on the term t. Namely (referring to the statements of $(\forall I)$ and $(\exists E)$ above), in $(\forall I)$, t is allowed to be a variable with no *free* occurrence in ϕ or undischarged assumptions of D, provided that t is substitutable for x in ϕ. Likewise in $(\exists E)$, t is allowed to be a variable with no free occurrence in χ, ϕ or any undischarged assumption of D' except $\phi[t/x]$, but again t must be substitutable for x in ϕ. These more general rules have a technical justification: if D is any derivation using them, then we can construct a derivation D' which has the same conclusion and undischarged assumptions as D, but is a σ-derivation in the sense we described earlier. This can be proved by a slight adjustment of the proof of Theorem 7.4.6. So we can count these more general rules as derived rules of our system of natural deduction.

On the other hand, informal mathematical arguments that correspond to these more general rules are generally frowned on. For example, if you start a proof by writing:

(7.35)   For every real number x, $x^2 + 1 \leqslant 2x$. Suppose x is nonzero but pure imaginary ...

your readers will try to read this as a donkey anaphora (compare (7.21)) and will wonder how a real number can be nonzero pure imaginary.

Exercises

7.4.1. (a) Show that asymmetric relations are always irreflexive; in other words, prove the following by natural deduction.

$$\{\forall x \forall y (R(x,y) \to \neg R(y,x))\} \vdash \forall x \neg R(x,x).$$

(b) Show that a function with a left inverse is injective; in other words, prove the following by natural deduction.

$$\{\forall x F(G(x)) = x\} \vdash \forall x \forall y (G(x) = G(y) \to x = y).$$

7.4.2. Give natural deduction derivations to prove the following sequents.

(a) $\{\forall x F(G(x)) = x\} \vdash \forall x \exists y\ x = F(y).$

(b) $\{\forall x \forall y \forall z\ ((R(x, y) \land R(y, z)) \to R(x, z))\} \vdash \forall x \forall y \forall z \forall w\ (R(x, y) \land R(y, z) \land R(z, w) \to R(x, w)).$

(c) $\{\neg \exists x P(x)\} \vdash \forall x \neg P(x).$

(d) $\{\neg \forall x \neg P(x)\} \vdash \exists x P(x)$ [Use (c)].

7.4.3. The following derivation proves the statement $2 \neq 3$
from the two assumptions $\forall x\ Sx \neq 0$ and $\forall x \forall y (Sx = Sy \to x = y)$. (In Example 7.7.3 we will learn that these assumptions are two of the axioms of Peano Arithmetic (PA).)

(a) Label each step with the name of the rule that it uses.

(b) Discharge any assumption that needs discharging, and show the step which cancels it.

(c) For each step of the form (=E), give the formula ϕ such that (for appropriate s, t and x) $\phi[s/x]$ is an assumption of the step and $\phi[t/x]$ is the conclusion.

(d) Rewrite the proof as a proof in ordinary mathematical English, but using the same steps where possible.

$$\frac{\dfrac{\dfrac{\forall x \forall y (Sx = Sy \to x = y)}{\forall y (SS0 = Sy \to S0 = y)}}{SS0 = SSS0 \to S0 = SS0}}{}$$

$$SS0 \neq SSS0$$

the corresponding semantic sequent has a counterexample; see Exercise 7.3.1(b).)

$$\cfrac{\dfrac{\forall x \exists y R(x,y)}{\exists y R(z,y)} \text{ (∀E)} \qquad \cfrac{\dfrac{\dfrac{\overline{R(z,u)}\,①}{\forall x R(x,u)} \text{ (∀I)}}{\exists y \forall x R(x,y)} \text{ (∃I)}}{}}{\underset{①}{\qquad\qquad\qquad} \exists y \forall x R(x,y)} \text{ (∃E)}$$

$$\exists y \forall x R(x,y)$$

7.5 Proof and truth in arithmetic

Definition 7.5.1 By a *diophantine formula* we mean a formula of $LR(\sigma_{\text{arith}})$ of the form

$$\exists y_1 \cdots \exists y_m \; (s = t).$$

A *diophantine sentence* is a diophantine formula that is a sentence.

Let n be a positive integer and X an n-ary relation on $\mathbb{N}$ (i.e. a set of n-tuples of natural numbers). It is clear that X is diophantine (as defined by Definition 5.8.1) if and only if there is a diophantine formula ϕ such that X is the set of all n-tuples that satisfy ϕ in $\mathbb{N}$. For example, the set of even natural numbers is the set of n such that (n) satisfies $\phi(x)$ in $\mathbb{N}$, where ϕ is $\exists y \; x = 2y$.

In this section we will examine what we can prove about diophantine formulas, using natural deduction. The following set of sentences is important for this:

Definition 7.5.2 We write PA_0 for the following set of sentences. (Example 7.7.3 will show the reason for this notation.)

(7.36)
$$
\begin{array}{ll}
(1) & \forall x \; (x + 0 = x) \\
(2) & \forall x \forall y \; (x + Sy = S(x+y)) \\
(3) & \forall x \; (x \cdot 0 = 0) \\
(4) & \forall x \forall y \; ((x \cdot Sy) = (x \cdot y) \; + \; x)
\end{array}
$$

Richard Dedekind Germany, 1831–1916. We can give recursive definitions for plus and times.

Theorem 7.5.3 (Dedekind's Theorem) *For every diophantine sentence ϕ,*

$$\models_N \phi \quad \Leftrightarrow \quad PA_0 \vdash_{\sigma_{arith}} \phi.$$

Proof Right to left follows from the fact that all the sentences in PA_0 are true in $\mathbb{N}$. (Here we are taking for granted that natural deduction has the Soundness property. We will prove it in the next section, without using Dedekind's Theorem.)

To prove left to right, we prove it first for sentences that are equations, by induction on the number of occurrences of function symbols $\bar{S}$, $\bar{+}$ or $\bar{\cdot}$ in the equations when they are written as equations of $LR(\sigma_{arith})$. There are several cases, according to the form of ϕ.

Case 1: ϕ is $k + m = n$ where $k, m, n \in \mathbb{N}$. If m is 0 then, since ϕ is true, k must be n and the sentence follows from (1) in PA_0 by $(\forall E)$. If m is Sm' then, since ϕ is true, n must be of the form Sn' and $m + n' = k'$. Now the equation $(\bar{+}(\bar{k}, \overline{m'}) = \overline{n'})$ has two fewer function symbols than $(\bar{+}(\bar{k}, \bar{m}) = \bar{n})$. So the induction hypothesis applies and tells us that $PA_0 \vdash k + m' = n'$. By (=term) this implies that $PA_0 \vdash S(k + m') = S(n')$. By (2) and $(\forall E)$, $PA_0 \vdash k + m = S(k + m')$. So by (=transitive), $PA_0 \vdash k + m = S(n')$. But n is $S(n')$, so $PA_0 \vdash k + m = n$.

Example 7.5.4 $PA_0 \vdash 2 + 2 = 4$. Exercise 5.4.4 was to explain the substitutions made in the applications of (=E) in the lower parts of this derivation. Here we have added applications of $(\forall E)$ to deduce the assumptions from PA_0.

$$\cfrac{\forall x \; x + 0 = x}{SS0 + 0 = SS0} \quad \cfrac{\cfrac{\forall x \forall y \; x + Sy = S(x + y)}{\forall y \; SS0 + Sy = S(SS0 + y)}}{SS0 + S0 = S(SS0 + 0)}$$

$$\cfrac{SS0 + S0 = SSS0}{SS0 + SS0 = SSSS0}$$

$$\cfrac{\cfrac{\forall x \forall y \; x + Sy = S(x + y)}{\forall y \; SS0 + Sy = S(SS0 + y)}}{SS0 + SS0 = S(SS0 + S0)}$$

Case 2: ϕ is $k \cdot m = n$ where $k, m, n \in \mathbb{N}$. The argument is like the previous case, using (3), (4) and the fact that Case One is proved.

Case 3: ϕ is $s + t = n$ or $s \cdot t = n$ or $St = n$ where $n \in \mathbb{N}$ and s, t are closed terms. First suppose ϕ is $s + t = n$ and is true in $\mathbb{N}$. The expressions s and t have number values p and q, and the equations $s = p$ and $t = q$ are true equations with fewer function symbols than ϕ (because the truth of ϕ implies that $p \leqslant n$ and $q \leqslant n$). So by induction assumption these two equations are provable from PA_0. Since ϕ is true, the sentence $p + q = n$ is true; hence by Case One, $PA_0 \vdash p + q = n$. We deduce $s + t = n$ by:

$$
\cfrac{
  t = q \qquad
  \cfrac{
    s = p \qquad \cfrac{}{s+t=s+t}\ \text{(=I)}
  }{s+t=p+t}\ \text{(=E)}
}{}
$$

$$
\cfrac{
  p + q = n \qquad
  \cfrac{
    t = q \qquad
    \cfrac{
      s = p \qquad s+t = s+t
    }{s+t = p+t}\ \text{(=E)}
  }{s+t = p+q}\ \text{(=E)}
}{s+t = n}\ \text{(=E)}
$$

When t has a value of at least 1, the case where ϕ is $s \cdot t = n$ is similar, using Case Two. When t has value 0, the equation $t = 0$ has fewer function symbols than $s \cdot t = 0$, so by induction hypothesis it is provable from PA_0, and then $s \cdot t = 0$ is provable using (3).

If ϕ is $St = n$, then since ϕ is true, n is Sq for some q. Then the equation $t = q$ is true and has fewer function symbols than $St = n$, so by induction hypothesis it is provable from PA_0; then $St = Sq$ is provable using (=term).

Case 4: ϕ is $s = t$ where s, t are closed terms. Let n be the number value of s. Then the equations $s = n$ and $t = n$ are true, and by the preceding cases they are provable from PA_0. So ϕ is provable from PA_0 by (=symmetric) and (=transitive).

This completes the proof when ϕ is quantifier-free. In the general case, suppose

$$\exists y_1 \cdots \exists y_n \psi$$

is true in $\mathbb{N}$, where ψ is an equation and the variables $y_1, \ldots, y_n$ are distinct. Then there are numbers $k_1, \ldots, k_n$ in $\mathbb{N}$ such that

$$\models_{\mathbb{N}} \psi[\bar{k}_1/y_1, \ldots, \bar{k}_n/y_n]$$

and hence

$$PA_0 \vdash \psi[\bar{k}_1/y_1, \ldots, \bar{k}_n/y_n]$$

by the quantifier-free case already proved.

We deduce that

$$PA_0 \vdash \exists y_1 \cdots \exists y_n \psi$$

as follows:

$$
\cfrac{
  \cfrac{
    \cfrac{\psi[\bar{k}_1/y_1, \ldots, \bar{k}_n/y_n]}{\exists y_n \psi[\bar{k}_1/y_1, \ldots, \bar{k}_{n-1}/y_{n-1}]}\ (\exists\text{I})
  }{\exists y_{n-1} \exists y_n \psi[\bar{k}_1/y_1, \ldots, \bar{k}_{n-2}/y_{n-2}]}\ (\exists\text{I})
}{\vdots}
$$

$$
\cfrac{\exists y_2 \cdots \exists y_n \psi[\bar{k}_1/y_1]}{\exists y_1 \cdots \exists y_n \psi}\ (\exists\text{I}) \qquad \square
$$

Exercises

7.5.1. Determine which of the following sentences are true in the structure $\mathbb{N}$. Give reasons for your answers.

(a) $\forall x \exists y (x = y + y \lor x = S(y + y))$.

(b) $\forall x \exists y \exists z (SSy \cdot SSz = SSSSx)$.

(c) $\forall x \forall y \exists z (SSy + SSz = SSSSx)$.

(d) $\forall x \exists y \exists z (x = y^2 + z^2)$.

(e) $\exists x \exists y \, (x \cdot Sy = x + Sy)$.

(f) $\exists x \exists y \, (x \cdot x = Sx + (y + y))$.

7.5.2. Describe a structure of the same signature as $\mathbb{N}$, but in which the following sentence (which is false in $\mathbb{N}$) is true:

$$\exists x \exists y \; SSS0 = x \cdot x + y \cdot y.$$

Justify your answer.

7.6 Soundness and completeness for first-order logic

Here we extend the results of Sections 5.9 and 5.10 from qf sentences to arbitrary sentences of LR. We say only what changes are needed.

We continue with the simplifying assumption that the truth function symbols are just $\land$, $\neg$ and $\bot$. We also assume that the only quantifier symbol is $\forall$. Theorem 7.3.7(a) guarantees that every formula of LR is logically equivalent to one in which the quantifier symbol $\exists$ never occurs.

Theorem 7.6.1 (Soundness of Natural Deduction for LR) *Let σ be a signature, Γ a set of sentences of $LR(\sigma)$ and ψ a sentence of $LR(\sigma)$. If $\Gamma \vdash_\sigma \psi$ then $\Gamma \models_\sigma \psi$.*

Proof We first assume that there is a σ-derivation D that proves $(\Gamma \vdash_\sigma \psi)$ and uses only sentences. Then we argue by induction on the height of D, as for Theorem 5.9.1 but with two new cases, viz. where the right label R on the bottom node of D is $(\forall I)$ or $(\forall E)$.

Case 2(f): R is $(\forall I)$ and the conclusion of D is $\forall x \phi$. Assuming x does occur free in ϕ, D must have the form

$$\frac{\begin{array}{c} D' \\ \phi[c/x] \end{array}}{\forall x \phi}$$

where D' is a derivation of ϕ from assumptions in Γ, and c is a constant symbol that does not occur in ϕ or any undischarged assumption of D'. Let $\Delta \subseteq \Gamma$ be the set of undischarged assumptions of D'; let A be a model of Γ and hence of Δ.

For each element a of A, write $A(a)$ for the structure got from A by changing the interpretation of c to be a, so that $c_{A(a)} = a$. Since c never appears in the sentences in Δ, each of the structures $A(a)$ is also a model of Δ by the Principle of Irrelevance (see before Lemma 7.3.3). So by induction hypothesis each $A(a)$ is a model of $\phi[c/x]$. But by Theorem 7.3.2 this means that each element a of A satisfies $\phi(x)$. Hence by Definition 7.3.1, A is a model of $\forall x \phi$.

If x does not occur free in ϕ then there is a more trivial argument using Theorem 7.3.7(g).

Case 2(g): R is ($\forall$E). Then D has the form

$$\frac{\begin{array}{c} D' \\ \forall x \phi \end{array}}{\phi[t/x]}$$

where all the undischarged assumptions of D' lie in Γ. Let A be a model of Γ. Then by induction hypothesis A is a model of $\forall x \phi$. If x does not occur in ϕ, then A is a model of ϕ by Theorem 7.3.7(g). If x does occur in ϕ, then t must be a closed term since $\phi[t/x]$ is a sentence; hence A has an element t_A. By Definition 7.3.1, since A is a model of $\forall x \phi$, every element of A satisfies ϕ. So in particular t_A satisfies ϕ, and hence by Theorem 7.3.2, A is a model of $\phi[t/x]$.

This proves the theorem under the assumption that D contains only sentences. To remove this assumption we take an infinite set W of witnesses, and we write σ^W for the signature got by adding these witnesses to σ. By Theorem 7.4.6 parts (a) (left to right) and (b), there is a σ^W-derivation which proves $(\Gamma \vdash \psi)$ and uses only sentences. By the case already proved, $\Gamma \models_{\sigma^W} \psi$, and so $\Gamma \models_\sigma \psi$ since Γ and ψ lie in LR(σ). $\square$

Theorem 7.6.2 (Adequacy of Natural Deduction for LR) *Let Γ be a set of sentences of LR(σ) and ψ a sentence of LR(σ). Then*

$$\Gamma \models_\sigma \psi \;\Rightarrow\; \Gamma \vdash_\sigma \psi.$$

As in Section 5.10, we prove this by showing that if a set of sentences Γ of LR(σ) is syntactically consistent (i.e. $\Gamma \nvdash_\sigma \bot$), then Γ can be extended to a Hintikka set, and every Hintikka set has a model.

Definition 7.6.3 We need to update Definition 5.10.2 to accommodate sentences with quantifiers. We say that a set Δ of sentences of LR(σ) is a *Hintikka set for RL(σ)* if it obeys conditions (1)–(7) of Definition 5.10.2 together with two new clauses:

(8) If $\forall x \phi$ is in Δ and t is a closed term of LR(σ), then $\phi[t/x]$ is in Δ.

(9) If $\neg \forall x \phi$ is in Δ then there is a constant symbol c of σ such that $\neg \phi[c/x]$ is in Δ.

Lemma 7.6.4 *If Δ is a Hintikka set for $LR(\sigma)$, then some σ-structure is a model of Δ.*

Proof We construct a σ-structure A from Δ exactly as in Lemma 5.10.3. Then we argue as before, using induction on complexity of sentences.

Besides the properties that we proved for A in the case of qf sentences, we have two more things to prove:

(7.37)   (a) If $\forall x\phi$ is in Δ then $\models_A \forall x\phi$.
           (b) If $\neg\forall x\phi$ is in Δ then $\forall x\phi$ is false in A.

To prove (a), if $\forall x\phi$ is in Δ then by clause (8), Δ contains $\phi[t/x]$ for every closed term t. So by induction hypothesis each sentence $\phi[t/x]$ is true in A, and hence every element t_A of A satisfies ϕ. (A subtle point: $\phi[t/x]$ could very well be a longer expression than $\forall x\phi$, if t is a large term. But we are going by induction on complexity of formulas—see Definition 5.3.9(a).) Since every element of A is named by a closed term, this proves that $\forall x\phi$ is true in A.

To prove (b), if $\neg\forall x\phi$ is in Δ then by clause (9), Δ contains $\neg\phi[c/x]$ for some constant symbol c. Since $\phi[c/x]$ has lower complexity than $\forall x\phi$, the induction hypothesis gives that $\phi[c/x]$ is false in A. So the element c_A fails to satisfy ϕ in A, and thus $\forall x\phi$ is false in A. $\qquad\square$

The next step is to show how to extend any syntactically consistent set to a Hintikka set, generalising Lemma 5.10.4. The new clause (9) gives us a headache here: in order to be sure that the set of sentences is still syntactically consistent after the sentence $\neg\phi[c/x]$ is added, we need to choose a constant symbol c that has not yet been used. Since the signature σ may not contain any spare constant symbols, we form a new signature σ^W by adding to σ an infinite set $W = \{c_0, c_1, \dots\}$ of witnesses.

Lemma 7.6.5 *Given a syntactically consistent set Γ of sentences of $LR(\sigma)$, there is a Hintikka set Δ for $LR(\sigma^W)$ with $\Gamma \subseteq \Delta$.*

Proof We note first that by Theorem 7.4.6(a), Γ is syntactically consistent in $LR(\sigma^W)$. We make a list $\theta_0, \theta_1, \dots$ which contains (infinitely often) each sentence of $LR(\sigma^W)$, and also contains (infinitely often) each triple $(\phi, x, s = t)$ as in the proof of Lemma 5.10.4. We also make sure that the list contains (infinitely often) each pair of the form $(\forall x\phi, t)$ where t is a closed term.

The sets Γ_i are constructed as before, but with two new cases. We check as we go that each Γ_i contains only finitely many witnesses from W. The set Γ_0 contained no witnesses, and each step will add just finitely many sentences and hence only finitely many new witnesses.

The two new cases are as follows.

- If θ_i is $(\forall x\phi, t)$ and $\forall x\phi$ is in Γ_i, then we put $\Gamma_{i+1} = \Gamma_i \cup \{\phi[t/x]\}$.

- If θ_i is $\neg\forall x\phi$ and $\theta_i \in \Gamma_i$, then let c be the first witness from W not used in Γ_i. (Here we use the induction hypothesis that Γ_i uses only finitely many of the witnesses.) We put $\Gamma_{i+1} = \Gamma_i \cup \{\neg\phi[c/x]\}$.

We have to show that if Γ_i is syntactically consistent in $\mathrm{LR}(\sigma^W)$, then so is Γ_{i+1}. We need only consider the two new cases above, since the argument proceeds as before in the other cases.

Case $(\forall x\phi, t)$. Suppose there is a derivation D of $\bot$ whose undischarged assumptions are in Γ_{i+1}. Then wherever $\phi[t/x]$ occurs as an undischarged assumption of D, extend the derivation upwards so as to derive the assumption from $\forall x\phi$ by $(\forall \mathrm{E})$. The result is a derivation proving $(\Gamma_i \vdash_{\sigma^W} \bot)$, which implies that Γ_i was already syntactically inconsistent.

Case $(\neg\forall x\phi)$. Suppose we have a derivation D of $\bot$ from Γ_{i+1}, that is, from Γ_i and $(\neg\phi[c/x])$. Since c never occurs in Γ_i, we can construct the derivation

$$(7.38) \qquad \begin{array}{c} \overbrace{(\neg\phi[c/x])}^{(1)} \\ D \\ \cfrac{\cfrac{\bot}{\phi[c/x]}\ (\mathrm{RAA})\ \scriptstyle(1)}{\forall x\phi} \ (\forall \mathrm{I}) \qquad\qquad (\neg\forall x\phi) \\ \hline \bot \end{array} \ (\neg\mathrm{E})$$

This is a derivation proving $(\Gamma_i \vdash_{\sigma^W} \bot)$, which shows that Γ_i was already syntactically inconsistent.

As in Lemma 5.10.4, we take Δ to be the union $\Gamma_0 \cup \Gamma_1 \cup \cdots$, and we confirm that Δ is a Hintikka set for $\mathrm{LR}(\sigma^W)$. $\qquad\square$

We can now state the following, for later use.

Theorem 7.6.6 *A set of sentences Γ of $LR(\sigma)$ is syntactically consistent if and only if it is consistent (i.e. has a model).*

Proof If Γ is syntactically consistent then we showed that Γ extends to a Hintikka set in an expanded signature σ^W, and that some σ^W-structure A is a model of Γ. The Principle of Irrelevance assures that A remains a model of Γ if we remove the interpretations of the symbols not in σ.

Conversely, assume Γ has a model. Then it is false that ($\Gamma \models \bot$), because if A is a model of Γ, all sentences of Γ are true in Γ, but $\bot$ is false. By Soundness, ($\Gamma \vdash \bot$) is false, that is, Γ is syntactically consistent. □

Arguing just as in Lemma 3.10.3, the Adequacy Theorem follows from Theorem 7.6.6. □

Combining Theorems 7.6.1 and 7.6.2 we get Theorem 7.6.7.

Theorem 7.6.7 (Completeness of Natural Deduction for LR) *If Γ is a set of sentences of $LR(\sigma)$ and ϕ is a sentence of $LR(\sigma)$, then*

$$\Gamma \models_\sigma \phi \;\Leftrightarrow\; \Gamma \vdash_\sigma \phi.$$

Theorem 7.6.8 (Compactness Theorem for LR) *If Γ is a set of sentences of LR, and every finite subset of Γ has a model, then Γ has a model.*

Proof Suppose Γ has no model. Then every model of Γ is a model of $\bot$, so $\Gamma \models \bot$. It follows by the Adequacy Theorem that there is a derivation D whose conclusion is $\bot$ and whose undischarged assumptions all lie in Γ. Let Γ' be the set of all undischarged assumptions of D. Since D is a finite object, Γ' is finite. Also since $\Gamma' \vdash \bot$, the Soundness Theorem tells us that $\Gamma' \models \bot$, and hence Γ' has no model. But we saw that $\Gamma' \subseteq \Gamma$. □

The Compactness Theorem is so-called because it can be interpreted as saying that a certain topological space associated to $LR(\sigma)$ is compact. It would be too great a digression to elaborate.

Exercises

7.6.1. By an *EA sentence* we mean a sentence of the form

$$\neg \forall y_1 \cdots \forall y_k \neg \forall z_1 \cdots \forall z_m \psi$$

where ψ is quantifier-free. Suppose ϕ is an EA sentence in a signature with no function symbols, and ℓ is the length of ϕ. Calculate in terms of ℓ an upper bound on the number of steps needed in Lemma 7.6.5 to construct a Hintikka set containing ϕ under the assumption that $\{\phi\}$ is syntactically consistent. [Thus $k \leqslant \ell$ steps are needed for applications of (9) in Definition 7.6.3, yielding a sentence with at most ℓ constants; then each of $m \leqslant \ell$ variables z_j needs to be replaced by each constant symbol in applications of (8); and so on.] Explain why it follows that if σ has no function symbols, then there is an algorithm for testing whether any given EA sentence of $LR(\sigma)$ is satisfiable.

7.7 First-order theories

First-order theories have two main uses. The first is to serve as definitions of classes of structures.

Example 7.7.1

(a) In Example 5.3.2(b) we introduced the signature σ_{group} for groups. The standard definition of groups can be written as a set of sentences in $\text{LR}(\sigma_{\text{group}})$, using universal quantifiers.

(7.39)
$$\forall x \forall y \forall z \; (x \cdot (y \cdot z) = (x \cdot y) \cdot z)$$
$$\forall x \; ((x \cdot e = x) \wedge (e \cdot x = x))$$
$$\forall x \; ((x \cdot x^{-1} = e) \wedge (x^{-1} \cdot x = e))$$

The class of groups is exactly the class of models of (7.39).

(b) In Example 5.3.2(c) we defined the signature of linear orders, σ_{lo}. The definition of a strict linear order (Definition 5.2.4) can be written using sentences of $\text{LR}(\sigma_{\text{lo}})$:

(7.40)
$$\forall x \; (\neg(x < x))$$
$$\forall x \forall y \forall z \; ((x < y) \wedge (y < z) \rightarrow (x < z))$$
$$\forall x \forall y \; (((x < y) \vee (y < x)) \vee (x = y))$$

By a *linearly ordered set* we mean a set together with a strict linear order on it. In other words, the class of linearly ordered sets is the class of all models of (7.40).

(c) The class of abelian groups is the class of models of the theory

(7.41)
$$\forall x \forall y \forall z \; (x + (y + z) = (x + y) + z)$$
$$\forall x \; (x + 0 = x)$$
$$\forall x \; (x + (-x) = 0)$$
$$\forall x \forall y \; (x + y = y + x)$$

We have not said what the signature is, but you can work it out from the sentences in the theory.

(d) Let K be a finite field. For each element k of K we introduce a 1-ary function symbol $\overline{k}$. Consider the theory consisting of (7.41) together with all sentences of the following forms:

(7.42)
$$\forall x \forall y \; (\overline{k}(x + y) = \overline{k}(x) + \overline{k}(y))$$
$$\forall x \; (\overline{k + h}(x) = \overline{k}(x) + \overline{h}(x))$$
$$\forall x \; (\overline{kh}(x) = \overline{k}(\overline{h}(x)))$$
$$\forall x \; (\overline{1}(x) = x)$$

for all elements k and h of K. The sentences (7.41) and (7.42) together express the usual definition of *vector spaces over* K. The reason for restricting to a finite field is that this allows us to have (at least in principle) a symbol for each element of the field. There are too many real numbers for us to symbolise them by expressions from a language with a finite lexicon, and it follows that we cannot define 'vector space over $\mathbb{R}$' by first-order axioms without a more abstract notion of 'symbol' than we have envisaged so far. We come back to this in Section 7.9.

Definition 7.7.2 Suppose Γ is a theory in $\mathrm{LR}(\sigma)$. The class of all σ-structures that are models of Γ is called the *model class* of Γ, and it is written $\mathrm{Mod}(\Gamma)$. We say also that Γ is a *set of axioms for* the class $\mathrm{Mod}(\Gamma)$.

In the terminology of Definition 7.7.2, each of the theories (7.39)–(7.41) is a set of axioms for a well-known class of mathematical structures, and the class is the model class of the theory. Of course, these three classes do not need formal sentences to define them. But first-order logic does provide a standard format for definitions of this kind. Also the fact that a class is a model class, with axioms of a particular form, often allows us to use methods of logic to prove mathematical facts about the structures in the class. We give a very brief sample in Section 7.9. (Also in Section 7.9 we comment on why Definition 7.7.2 speaks of 'model classes' rather than 'model sets'.)

The second main use of first-order theories is to record a set of facts about a particular structure. For example, here is a theory that states properties of the natural number structure $\mathbb{N}$.

Example 7.7.3 The σ_{arith}-structure $\mathbb{N}$ is a model of the following infinite set of sentences:

(1) $\forall x\ (0 \neq Sx)$.

(2) $\forall x \forall y\ (Sx = Sy \to x = y)$.

(3) $\forall x\ (x + 0 = x)$.

(4) $\forall x \forall y\ (x + Sy = S(x + y))$.

(5) $\forall x\ (x \cdot 0 = 0)$.

(6) $\forall x \forall y\ (x \cdot Sy = (x \cdot y)\ +\ x)$.

(7) For every formula ϕ with $FV(\phi) = \{x, y_1, \ldots, y_n\}$,
     the sentence $\forall y_1 \ldots \forall y_n\ ((\phi[0/x] \wedge \forall x(\phi \to \phi[Sx/x])) \to \forall x \phi)$.

This set of sentences is called *first-order Peano Arithmetic*, or PA for short. You recognise (3)–(6) as the sentences of PA_0 in Definition 7.5.2.

Here (7) is as near as we can get within first-order logic to an axiom that Giuseppe Peano wrote down in 1889. In place of (7) he had a single sentence expressing:

(7.43)   If X is any set that contains 0, and is such that {if a natural number n is in X then so is $n+1$}, then all natural numbers are in X.

We cannot express (7.43) as a first-order sentence about $\mathbb{N}$, because it needs a universal quantifier $\forall X$ where X is a variable for sets of numbers, not single numbers. Languages with quantifiers ranging over sets of elements, or over relations or functions on the domain, are said to be *second-order*. We can use them to say more than we could express in first-order logic, but at a cost: various valuable facts about first-order logic, such as the existence of proof calculi with the Completeness property, are not true for second-order logic.

Giuseppe Peano Italy, 1858–1932. Logic and arithmetic can be reduced to a few primitive notions.

Example 7.7.4 Here is an important notion that can be expressed in second-order logic; we will see in Exercise 7.9.3 that it cannot be expressed in first-order logic. A *well-ordered set* is a model of the following set of sentences of $\mathrm{LR}(\sigma_{\mathrm{lo}})$, and its order relation $<_A$ is called a *well-order*. The quantifier $\forall X$ ranges over sets of individuals, and $X(y)$ means that y is in the set X.

(1) $\forall x\ (\neg(x < x))$.

(2) $\forall x\ \forall y \forall z((x < y \wedge y < z) \to (x < z))$.

(3) $\forall x\ \forall y((x < y) \vee (x = y) \vee (y < x))$.

(4) $\forall X\ (\exists y X(y) \to \exists y\ (X(y) \wedge \forall z(z < y \to \neg X(z))))$.

Sentences (1)–(3) are (7.40), the axioms for a linearly ordered set. Sentence (4) says that every non-empty set of elements of the ordered set has a least element. So, for example, the natural numbers with their usual linear order form a well-ordered set. But the real numbers with their usual linear order do not, because there is no least real number. Also the set of non-negative real numbers is not a well-ordered set; it has a least element, but the subset consisting of the positive real numbers is non-empty and has no least element.

Complete theories

Definition 7.7.5 Let Γ be a theory in $LR(\sigma)$.

(a)   A *consequence* of Γ is a sentence ϕ of $LR(\sigma)$ such that $\Gamma \models \phi$. We write $\overline{\Gamma}$ for the set of consequences of Γ.

(b)   The theory Γ is called *complete* if for every sentence ϕ of $LR(\sigma)$, exactly one of ϕ and $(\neg\phi)$ is a consequence of Γ.

Definition 7.7.6

(a)   If A is a σ-structure, the *theory of* A, written $\mathrm{Th}(A)$, is the set of all sentences ϕ of $LR(\sigma)$ that are true in A. (So A is a model of $\mathrm{Th}(A)$.)

(b)   If A and B are σ-structures, we say that A and B are *elementarily equivalent* (written $A \equiv B$) if $\mathrm{Th}(A) = \mathrm{Th}(B)$. Then $\equiv$ is an equivalence relation on the class of σ-structures; its equivalence classes are called *elementary equivalence classes*. (So each σ-structure A belongs to one equivalence class, namely the class of all σ-structures elementarily equivalent to A.)

Theorem 7.7.7 *Let Γ be a theory in $LR(\sigma)$. The following are equivalent:*

(a) Γ is complete.

(b) Γ is consistent (Definition 7.3.4(d)), and for every sentence ϕ of $LR(\sigma)$, at least one of ϕ and $(\neg\phi)$ is a consequence of Γ.

(c) Γ is consistent, and if A, B are models of Γ then $A \equiv B$.

(d) $\overline{\Gamma} = Th(A)$ for some σ-structure A.

Proof   **(a)** $\Rightarrow$ **(b)**: Suppose Γ is not consistent. Then Γ has no models, and so trivially every model of Γ is a model of both $\bot$ and $(\neg\bot)$, proving that both $\bot$ and $(\neg\bot)$ are consequences of Γ. So (a) implies that Γ is consistent. The second part of (b) is immediate from (a).

   (b) $\Rightarrow$ **(c)**: Assume (b) and let A, B be models of Γ. If ϕ is a consequence of Γ then ϕ is in both $\mathrm{Th}(A)$ and $\mathrm{Th}(B)$; if $(\neg\phi)$ is a consequence of Γ then $(\neg\phi)$ is in both $\mathrm{Th}(A)$ and $\mathrm{Th}(B)$. By (b), one or other of these alternatives holds for every sentence ϕ of $LR(\sigma)$. Since $\mathrm{Th}(A)$ and $\mathrm{Th}(B)$ are clearly both consistent, neither of them contains both ϕ and $(\neg\phi)$ for any sentence ϕ, and so we infer that $\mathrm{Th}(A) = \mathrm{Th}(B)$.

   (c) $\Rightarrow$ **(d)**: Assume (c). Then Γ has a model A; we claim that $\overline{\Gamma} = \mathrm{Th}(A)$. If ϕ is a consequence of Γ then $\models_A \phi$ since A is a model of Γ. If $\phi \in \mathrm{Th}(A)$ then by (c), $\phi \in \mathrm{Th}(B)$ for every model B of Γ, and so ϕ is a consequence of Γ.

(d) ⇒ (a): Assume (d) and suppose $\overline{\Gamma} = \text{Th}(A)$. Then for every sentence ϕ of $\text{LR}(\sigma)$, ϕ is a consequence of Γ if and only if ϕ is true in A. But for every sentence ϕ of $\text{LR}(\sigma)$, exactly one of ϕ and $(\neg\phi)$ is true in A.   □

Exercises

7.7.1. Construct a suitable signature for talking about abelian groups.

7.7.2. We write σ_{equiv} for the signature consisting of one binary relation symbol E. In the language $\text{LR}(\sigma_{\text{equiv}})$, write a theory expressing that E is an equivalence relation, so that a σ_{equiv}-structure A is a model of your theory if and only if the relation E_A is an equivalence relation on the domain of A.

7.7.3. The *finite spectrum* of a theory Γ is the set $\{n \in \mathbb{N} \mid \Gamma$ has a model with exactly n elements$\}$.

   (a) Write a first-order theory whose finite spectrum is the set of positive multiples of 3.

   (b) Write a first-order theory whose finite spectrum is the set of all positive integers $\equiv 1 \pmod 3$.

   (c) Write a first-order theory whose finite spectrum is the set of positive integers of the form n^2.

   (d) Write a first-order theory whose finite spectrum is the set of all prime powers > 1.

   (e) Write a first-order theory whose finite spectrum is the set of all positive integers of the form 5^n.

   (An old unsolved problem of Heinrich Scholz asks whether for every finite spectrum X of a finite first-order theory, the set $\mathbb{N} \setminus (X \cup \{0\})$ is also the finite spectrum of a finite first-order theory.)

7.7.4. Consider Example 7.7.1(b), the theory of linear ordered sets.

   (a) A linearly ordered set is called *discrete without endpoints* if for every element a there are a greatest element $< a$ and a least element $> a$. Write a sentence ϕ so that (7.40) together with ϕ are a set of axioms for the class of discrete linearly ordered sets without endpoints.

   (b) Describe all discrete linearly ordered sets without endpoints. [If A is a discrete linearly ordered set without endpoints, define an equivalence relation $\sim$ on the domain of A by: $a \sim b$ if and only if there are at most finitely many elements between a and b. What can you say about the equivalence classes?]

7.8 Cardinality

We can write down a name for each natural number. More precisely, if n is a natural number then, given enough time and paper, we could write down the arabic numeral for n. Our language for arithmetic, $\mathrm{LR}(\sigma_{\mathrm{arith}})$, contains a name $\bar{n}$ for each natural number n, and in several places we have used this fact. Can one write down a name for each real number?

The answer is no. Again more precisely, there is no system of expressions for all real numbers so that if r is any real number, then given enough time and paper, we could eventually write down the expression for r. This was proved by Georg Cantor in the nineteenth century. He showed that there are too many real numbers; the number of real numbers is greater than the number of expressions of any language with a finite lexicon. This fact is a theorem of *set theory*, a subject that Cantor created. Set theory not only tells us that we cannot name each real number by a written expression; it also tells us where to find collections of 'abstract' symbols of any size we want. But this is not a course in set theory, so here we will mostly confine ourselves to the facts needed to explain and prove the result of Cantor just mentioned.

According to Cantor, the set X has the same size as the set Y (in symbols $X \approx Y$) if and only if there is a binary relation R consisting of ordered pairs (x, y) with $x \in X$ and $y \in Y$, such that each $x \in X$ is the first term of exactly one pair in R and each $y \in Y$ is the second term of exactly one pair in R. A relation of this form is known as a *one-to-one correspondence* between X and Y. By Definition 5.2.7 such a relation R is in fact a function from X to Y (taking x to y if and only if the pair (x, y) is in R). We will refer to a function of this form as a *bijective function* from X to Y, or more briefly a *bijection* from X to Y. The relation $\approx$ is an equivalence relation on the class of all sets. For example, the identity function from X to X shows that $X \approx X$; see Exercise 7.8.1.

Cantor also proposed that we count the set X as having smaller size than the set Y (in symbols $X \prec Y$) if X and Y do not have the same size but there is an injective function $g : X \to Y$.

At first sight it seems that Cantor must have made a mistake, because there are pairs of sets that are in the relation $\approx$ but clearly have different sizes. For example, the function

$$\tan^{-1} \ : \ \left(-\frac{\pi}{2}, \frac{\pi}{2} \right) \to \mathbb{R}$$

is a bijection from the bounded interval $(-\pi/2, \pi/2)$ to the set of all real numbers. Fortunately, Cantor was not distracted by examples like this. He took the sensible view that there are different kinds of size. The kind of size that is measured by $\approx$ is called *cardinality*. Today Cantor's definitions of 'same size' and 'smaller size' in terms of $\approx$ and $\prec$ are widely accepted.

Another objection that Cantor had to meet is that his definition of 'same cardinality' does not answer the question 'What is a cardinality?' We do not propose to give an answer to this question; every available answer uses a serious amount of set theory. But we can set down here what we are going to assume. For a start, we will assume that each set X has a cardinality $\text{card}(X)$. The objects $\text{card}(X)$ that are used to represent cardinalities are known as *cardinals*.

For finite sets the natural numbers serve as cardinals. We say that a set X is *finite* if there is a natural number n such that X can be listed without repetition as

$$(7.44) \qquad\qquad X = \{x_0, \ldots, x_{n-1}\}$$

and in this case we count the number n (which is unique) as the cardinality of X. The function $i \mapsto x_i$ is a bijection from $\{0, \ldots, n-1\}$ to X. Also if X and Y are finite sets then $X \prec Y$ if and only if $\text{card}(X) < \text{card}(Y)$. The empty set is the only set with cardinality 0. We say that a set is *infinite* if it is not finite.

According to standard axioms of set theory, the cardinals are linearly ordered by size, starting with the finite cardinalities (the natural numbers) and then moving to the infinite ones. We have

$$\text{card}(X) = \text{card}(Y) \text{ if and only if } X \approx Y;$$
$$\text{card}(X) < \text{card}(Y) \text{ if and only if } X \prec Y.$$

So by the definitions of $\approx$ and $\prec$, $\text{card}(X) \leqslant \text{card}(Y)$ if and only if there is an injective function $f : X \to Y$.

Also the cardinals are well-ordered: every non-empty set of cardinals has a least member. We can deduce from this that if there are any infinite cardinals then there is a least one; it is known as ω_0. If there are any cardinals greater than ω_0, then there is a least one, known as ω_1; and so on. So the ordering of the cardinals starts like this:

$$0, 1, 2, \ldots, \omega_0, \omega_1, \omega_2, \ldots$$

at least if infinite cardinals exist. But they do: the set $\mathbb{N}$ is infinite, so its cardinality $\text{card}(\mathbb{N})$ is an infinite cardinal. Again according to standard axioms of set theory, $\text{card}(\mathbb{N}) = \omega_0$; the set of natural numbers is as small as an infinite set can be, reckoned by cardinality. Also a theorem of Cantor states that there is no greatest cardinal (Exercise 7.8.7); so ω_1 must exist too, and likewise ω_2 and onwards.

We begin our detailed treatment by discussing the sets of cardinality ω_0. Then we will prove Cantor's Theorem stating that the cardinality of $\mathbb{R}$ is greater than ω_0.

Definition 7.8.1

(a) A set X is *countably infinite* if $X \approx \mathbb{N}$; in other words, X is countably infinite if it can be listed without repetition as

(7.45) $$X = \{x_0, x_1, \ldots\}$$

where the indices 0, 1 etc. run through the natural numbers. (So the countably infinite sets are those of cardinality ω_0.)

(b) A set X is *countable* (or *enumerable*) if it is either finite or countably infinite; in other words, X is countable if it can be listed without repetition as in (7.45) where the list may or may not go on for ever. A set is said to be *uncountable* if it is not countable. (So all uncountable sets are infinite.)

In this definition, 'without repetition' means that if $i \neq j$ then $x_i \neq x_j$. Listing a set of things without repetition, using natural numbers as indices, is the main thing we do when we count the set; this is the idea behind the word 'countable'. Note also that if $f : X \to Y$ is bijective and X is listed without repetition as $\{x_0, x_1, \ldots\}$, then $\{f(x_0), f(x_1), \ldots\}$ is a corresponding listing of Y. So if X is countable and $Y \approx X$ then Y is countable too (and likewise with finite and countably infinite).

Lemma 7.8.2 *Let X be a set. The following are equivalent:*

(a) X is countable.

(b) X is empty or has an infinite listing as $\{x_0, x_1, \ldots\}$, possibly with repetitions.

(c) There is an injective function $f : X \to \mathbb{N}$.

Proof (a) $\Rightarrow$ (b): If X is finite and not empty, say $X = \{x_1, \ldots, x_n\}$, then we can list X as $\{x_0, \ldots, x_n, x_n, x_n, \ldots\}$.

(b) $\Rightarrow$ (c): If X is empty then the empty function from X to $\mathbb{N}$ is injective. (The empty function f takes nothing to nothing; so in particular there are no distinct x and y with $f(x) = f(y)$.) If X is not empty then, given the listing of X, define a function $h : X \to \mathbb{N}$ by

$$h(x) = \text{ the least } n \text{ such that } x = x_n.$$

Then h is an injective function from X to $\mathbb{N}$.

(c) $\Rightarrow$ (a): First suppose that Y is any subset of $\mathbb{N}$. Then Y can be listed in increasing order, say as

(7.46) $$Y = \{y_0, y_1, \ldots\}.$$

How do we know this? Using the well-order of $\mathbb{N}$ (Example 7.7.4), take y_0 to be the least element of Y, y_1 the least element of $Y \setminus \{y_0\}$, y_2 the least element of $Y \setminus \{y_0, y_1\}$ and so on. There is no harm if after some y_n there are no elements of Y left; this just means Y is finite. We need to check that every element of

Y appears eventually in the list. If not, then by well-ordering there is a least element y that is not listed. But by the way the list elements are chosen, y will be the least element of $Y \setminus \{y_0, \ldots, y_k\}$ as soon as we have listed among $y_0, \ldots, y_k$ all the elements of Y that are less than y; so y will in fact be listed after all.

Now assume (c), and take Y to be the image $\{f(x) \mid x \in X\}$ of X in $\mathbb{N}$. Then using (7.46) we can list X without repetition as

$$X = \{f^{-1}(y_0), f^{-1}(y_1), \ldots\}$$

so that X is countable too. □

Next we show that various operations on countable sets lead to countable sets. If X and Y are sets, we write $X \times Y$ for the set of all ordered pairs (x, y) where $x \in X$ and $y \in Y$; $X \times Y$ is the *cartesian product* of X and Y.

Theorem 7.8.3

(a) *A subset of a countable set is countable. More generally if $f : X \to Y$ is an injective function and Y is countable then X is countable too.*

(b) *The cartesian product of two countable sets X, Y is countable.*

(c) *If $f : X \to Y$ is surjective and X is countable then Y is countable.*

(d) *The union of two countable sets is countable.*

Proof (a) Suppose Y is countable and $f : X \to Y$ is injective. Then by Lemma 7.8.2 there is an injective function $g : Y \longrightarrow \mathbb{N}$, and so the composite function $g \circ f$ is an injective function from X to $\mathbb{N}$. Hence Y is countable by Lemma 7.8.2 again. (If X is a subset of Y, take f to be the identity function on X.)

(b) Suppose X and Y are countable. Then their elements can be listed without repetition as $X = \{x_0, x_1, \ldots\}$ and $Y = \{y_0, y_1, \ldots\}$. We define a function $F : X \times Y \to \mathbb{N}$ by

$$F(x_i, y_j) = 2^i 3^j$$

By unique prime decomposition F is injective. Hence $X \times Y$ is countable.

(c) Suppose $f : X \to Y$ is surjective. If X is empty then so is Y. Suppose X is non-empty and countable, so that by Lemma 7.8.2 we can list the elements of X as an infinite list $\{x_0, x_1, \ldots\}$. Then $\{f(x_0), f(x_1), \ldots\}$ is a listing of Y. It follows that Y is countable by Lemma 7.8.2 again.

(d) If either X or Y is empty the result is obvious; so assume neither is empty. Since X is countable, by Lemma 7.8.2 it has an infinite listing as $\{x_0, x_1, \ldots\}$. Likewise Y has an infinite listing $\{y_0, y_1, \ldots\}$. Now

$$\{x_0, y_0, x_1, y_1, \ldots\}$$

lists $X \cup Y$, so $X \cup Y$ is countable. □

These and similar arguments show that various sets are countable. For example the set of positive integers, the set of squares and the set of integers are all countably infinite (Exercise 7.8.2). The next example illustrates all the parts of Theorem 7.8.3.

Corollary 7.8.4 *The set $\mathbb{Q}$ of all rational numbers is countable.*

Proof By Theorem 7.8.3(a,b) the set X of ordered pairs (m, n) of natural numbers with $n \neq 0$ is countable. There is a surjective function $(m, n) \mapsto m/n$ from X to the set Y of non-negative rational numbers, and so Y is countable by Theorem 7.8.3(c). The bijection $x \mapsto -x$ shows that the set Z of non-positive rational numbers is countable too. Then $\mathbb{Q} = Y \cup Z$ is countable by Theorem 7.8.3(d). $\square$

Georg Cantor Germany, 1845–1918. Founder of set theory and the arithmetic of infinite numbers.

By now you may be wondering whether all infinite sets are countable. The famous Cantor's Theorem shows that this is not the case. The full theorem is Exercise 7.8.7; the following is an important special case.

Theorem 7.8.5 *Let m and w be two distinct objects, and let F be the set of all functions $f : \mathbb{N} \to \{m, w\}$. Then F is uncountable.*

Proof Suppose $\{f_0, f_1, \dots\}$ is an infinite listing of elements of F. We show that the listing does not list the whole of F, by finding an element g that is not included. The element g is defined as follows: For each natural number n,

$$(7.47) \qquad g(n) = \begin{cases} m & \text{if } f_n(n) = w \\ w & \text{if } f_n(n) = m \end{cases}$$

Then for each n, $g \neq f_n$ since $g(n) \neq f_n(n)$. This shows that F is not countable. $\square$

A historical curiosity is that Cantor seems to have chosen m and w to stand for man and woman (Mann and Weib in German)! His proof is known as the *diagonal argument*, because we can imagine forming g by choosing the opposites of the elements in the NW/SE diagonal of the infinite matrix

$$
\begin{array}{ccccc}
f_0(0) & f_0(1) & f_0(2) & f_0(3) & \cdots \\
f_1(0) & f_1(1) & f_1(2) & f_1(3) & \cdots \\
f_2(0) & f_2(1) & f_2(2) & f_2(3) & \cdots \\
f_3(0) & f_3(1) & f_3(2) & f_1(3) & \cdots \\
\vdots & \vdots & \vdots & \vdots &
\end{array}
$$

We can now exhibit another uncountable set.

Corollary 7.8.6 *The set $\mathbb{R}$ of all real numbers is uncountable.*

Proof   In Theorem 7.8.5 choose m and w to be 5 and 6. Then each function f in F describes a real number with decimal expansion

$$0 \, . \, f(0) \, f(1) \, f(2) \, \cdots$$

consisting entirely of 5s and 6s. This gives an injective function $h : F \to \mathbb{R}$. Since F is not countable, neither is $\mathbb{R}$, by Theorem 7.8.3(a). $\square$

By Corollary 7.8.6, card($\mathbb{R}$) is at least ω_1. The question whether card($\mathbb{R}$) is exactly ω_1 is known as the *continuum problem* (referring to the real numbers as the continuum). We know that the set-theoretic axioms that are commonly assumed today are not strong enough to answer this question.

The more general version of Theorem 7.8.5 in Exercise 7.8.7 implies that for every set X there is a set with greater cardinality than X. We get into trouble if we take X to be the whole universe of mathematics, since there is nothing greater than the universe. This is why today we distinguish between *sets* (which have cardinalities) and *proper classes* (which are classes that are too large to have cardinalities). The class of all sets is a proper class, and so is the class of all cardinals. If X is a proper class and $X \subseteq Y$ then Y is a proper class too.

Although we are barred from speaking of the class of *all* subclasses of a proper class X, there is nothing to prevent us defining and using particular subclasses of X. Likewise we can introduce binary relations on X; already in this section we have discussed the equivalence relation $\approx$ on the proper class of all sets, and Section 7.9 will discuss some other equivalence relations on proper classes.

Exercises

7.8.1.   Show that the relation $\approx$ is an equivalence relation on the class of sets. [For reflexive, use the identity function from a set to itself. For symmetric, use the inverse of a bijection. For transitive, use the composition of two bijections.]

7.8.2. (a) Write down a bijection from the interval $[0, 1]$ to the interval $[0, 2]$ in $\mathbb{R}$.

(b) Write down a bijection from the set $\mathbb{N}$ to the set of positive integers.

(c) (Galileo's Paradox) Write down a bijection from $\mathbb{N}$ to the set of squares of integers.

(d) Describe a bijection from $\mathbb{N}$ to the set $\mathbb{Z}$ of integers.

(e) Describe a bijection from the closed interval $[0, 2\pi]$ in $\mathbb{R}$ to the set of points of the unit circle $x^2 + y^2 = 1$.

7.8.3. Let $X_0, X_1, X_2, \ldots$ be countable sets (the list may or may not terminate). Prove that the union $X_0 \cup X_1 \cup X_2 \cdots$ is countable. (This is usually expressed by saying 'a countable union of countable sets is countable'.)

7.8.4. (a) Prove that the set S of all finite sequences of natural numbers is countable. [Define $F : S \to \mathbb{N}$ by

$$F(k_1, \ldots, k_n) = 2^{k_1+1} \cdot 3^{k_2+1} \cdot \ldots \cdot p_n^{k_n+1}$$

where p_n is the n-th prime.]

(b) Deduce that the set of all finite sets of natural numbers is countable.

(c) Show that if σ is a first-order signature with countably many symbols, then the set of terms of $\mathrm{LR}(\sigma)$ is countable and the set of formulas of $\mathrm{LR}(\sigma)$ is countably infinite.

7.8.5. The *algebraic numbers* are the complex numbers that satisfy some polynomial equation with integer coefficients. Prove that the set of algebraic numbers is countable.

7.8.6. The *Unrestricted Comprehension Axiom* of set theory states that if $P(x)$ is any predicate, there is a set $\{a \mid P(a)\}$ consisting of all objects a such that $P(a)$ is true. Show that if $P(x)$ is the predicate '$x \notin x$' and $b = \{a \mid a \notin a\}$, then $b \in b$ if and only if $b \notin b$. (This is the Russell–Zermelo paradox, which has led most set theorists to reject the Unrestricted Comprehension Axiom.)

7.8.7. Let X be a set; we write $\mathcal{P}X$ for the set of all subsets of X.
(a) Show that $\mathrm{card}(X) \leqslant \mathrm{card}(\mathcal{P}X)$. [Take x to $\{x\}$.]

(b) (Cantor's Theorem) Prove that $\mathrm{card}(X) < \mathrm{card}(\mathcal{P}X)$. [By (a) it suffices to show that the cardinalities are not equal; in fact we show that no function $h : X \to \mathcal{P}X$ is surjective. Consider $Y = \{x \in X \mid x \notin h(x)\}$, and show that Y must be different from each set in the image of h.]

7.8.8. The purpose of this exercise is to show that Exercise 7.8.7 generalises the diagonal argument in Theorem 7.8.5. For every subset Z of X the *characteristic function* χ_Z is defined by

$$\chi_Z(x) = \begin{cases} 1 & \text{if } x \in Z \\ 0 & \text{if } x \notin Z \end{cases}$$

Writing F for the set of all functions from X to $\{0,1\}$, show that the function $Z \mapsto \chi_Z$ is a bijection from $\mathcal{P}X$ to F, so that $\mathcal{P}X \approx F$. Show that the set Y in Exercise 7.8.7 is defined by putting, for each x,

$$\chi_Y(x) = 1 - \chi_{h(x)}(x).$$

So if $X = \mathbb{N}$ and $f_n = \chi_{h(n)}$ then χ_Y is the diagonal function g of the proof of Theorem 7.8.5 (with 1 for m and 0 for w).

7.9 Things that first-order logic cannot do

How far can you pin down a mathematical structure A by giving its first-order theory $\mathrm{Th}(A)$?

In a sense, you cannot at all. For example, take the natural number structure $\mathbb{N}$, and replace the number 0 by Genghis Khan. The arithmetic is just the same as before, except that now $2 - 2$ is Genghis Khan, Genghis Khan times 2 is Genghis Khan, Genghis Khan plus 3 is 3 and so on. This just goes to show that for mathematical purposes the identity of the elements of a structure is generally irrelevant; all that interests us is the relationships between the elements in the structure. To make this thought precise, we introduce the idea of an isomorphism, which is a translation from one structure to another that preserves all the mathematically interesting features.

Definition 7.9.1 Let A, B be σ-structures. By an *isomorphism* from A to B we mean a bijection f from the domain of A to the domain of B such that

(1) For all constant symbols c of σ,

$$f(c_A) = c_B$$

(2) If F is a function symbol of σ with arity n, and $a_1, \ldots, a_n$ are elements of A, then

$$f(F_A(a_1, \ldots, a_n)) = F_B(f(a_1), \ldots, f(a_n))$$

(3) If R is a relation symbol of σ with arity n, and $a_1, \ldots, a_n$ are elements of A, then

$$R_A(a_1, \ldots, a_n) \Leftrightarrow R_B(f(a_1), \ldots, f(a_n))$$

We say that A is *isomorphic to B*, in symbols $A \cong B$, if there is an isomorphism from A to B.

For example, if A and B are groups (taken as σ_{group}-structures, Example 5.3.2(b)), then Definition 7.9.1 agrees with the usual notion of group isomorphism. In Exercises 5.5.6 and 5.5.7 we saw some examples of isomorphisms between digraphs. If σ_{lo} is the signature of linear orders (as in Example 5.3.2(c)) and the σ_{lo}-structures A, B are linearly ordered sets with order relations $<_A$ and $<_B$, then an isomorphism f from A to B is called an *order-isomorphism*; it is a bijective function from the domain of A to the domain of B, such that for all a, b in the domain of A, $a <_A b$ implies $f(a) <_B f(b)$.

The relation $\cong$ is an equivalence relation on the class of all σ-structures, and its equivalence classes are known as *isomorphism classes*. As our argument with Genghis Khan shows, each isomorphism class is as large as the whole universe, because for any structure A and any object b not in A we can construct an isomorphic copy of A that has b as an element. In the terminology of Section 7.8, isomorphism classes are always proper classes.

Theorem 7.9.2 *Let A and B be isomorphic σ-structures. Then $A \equiv B$.*

This is intuitively obvious, since A and B are mathematically speaking identical. But one can also give a rigorous proof by induction on the complexity of formulas of $\mathrm{LR}(\sigma)$. Theorem 7.9.2 tells us that an elementary equivalence class of σ-structures is the union of one or more isomorphism classes. It is not hard to see that every non-empty model class is the union of one or more elementary equivalence classes; so non-empty model classes are always proper classes too. This is why mathematicians talk about the *class* of all groups, not about the set of all groups.

So we cannot identify a particular structure A by giving its first-order theory $\mathrm{Th}(A)$. But can $\mathrm{Th}(A)$ identify the isomorphism class of A? In other words, does the converse of Theorem 7.9.2 hold?

The answer is known to be Yes if A has finitely many elements: any other model of $\mathrm{Th}(A)$ must be isomorphic to A. But if A has infinitely many elements, the answer is always No; there are structures elementarily equivalent to A but not isomorphic to it. Here is the reason why.

Definition 7.9.3

(a) We define the *cardinality* of a structure to be the cardinality of its domain. The *spectrum* of a theory Γ is the class of cardinals κ such that Γ has a model of cardinality κ.

(b) We define the *cardinality* of a language L to be the cardinality of the set of sentences of L. (One can show that the cardinality of $\mathrm{LR}(\sigma)$ is the least infinite cardinal $\geqslant \mathrm{card}(\sigma)$.)

For the languages that we have considered so far in our general theory, the cardinality is always ω_0 (by Exercise 7.8.4). But one can give a more abstract definition that allows the symbols in signatures to be any mathematical objects, not necessarily shapes that can be written down; in this more general setting the cardinality of $LR(\sigma)$ might be any infinite cardinal. The theorems of Section 7.6—in particular, the Completeness Theorem and the Compactness Theorem— remain true after this generalisation, though the proofs need to be adjusted to use some devices of set theory.

The following theorem extracts parts of two other more detailed theorems known as the *Upward and Downward Löwenheim-Skolem Theorems*.

Theorem 7.9.4 (Spectrum Theorem) *Suppose $LR(\sigma)$ has cardinality κ and Γ is a theory in $LR(\sigma)$.*

(a) *If Γ has a model with infinite cardinality then every cardinal $\geq \kappa$ is in the spectrum of Γ.*

(b) *If the spectrum of Γ contains infinitely many natural numbers, then Γ has a model with infinite cardinality.*

Proof   We sketch the proof. It uses two main devices: the construction of models of Hintikka sets in the proof of Lemma 5.10.3, and the Compactness Theorem (Theorem 7.6.8).

Consider the models of Hintikka sets first. If Δ is a Hintikka set for $LR(\tau)$, then the domain of a model A of Δ consists of the equivalence classes $t^{\sim}$ of closed terms t of $LR(\tau)$. One can prove from this that the cardinality of A is at most that of the set of closed terms of $LR(\tau)$, and this in turn is at most the cardinality of $LR(\tau)$. (When τ is countable, $LR(\tau)$ has countably many closed terms—see Exercise 7.8.4(c). It follows by Theorem 7.8.3(c) that A is countable.)

Suppose λ is any cardinal $\geq \kappa$. Take a set W of witnesses, of cardinality λ. (Here we use an abstract set-theoretic notion of symbols.) Write σ^W for σ with the symbols of W added. Let Γ' be Γ together with all the sentences

(7.48) $(c \neq d)$ for all distinct witnesses c, d in W

If Δ is a Hintikka set for $LR(\sigma^W)$ with $\Gamma' \subseteq \Delta$, and A is the model of Δ constructed in the proof of Lemma 5.10.3, then it can be shown that $\operatorname{card}(A) \leq \operatorname{card}(LR(\sigma^W)) = \lambda$. But if c, d are distinct   witnesses from W, then $(c = d)$ cannot be in Δ because of (7.48), and so by (5.58) we have $c^{\sim} \neq d^{\sim}$. It follows that the function $c \mapsto c^{\sim}$ from W to the domain of A is injective, hence $\lambda \leq \operatorname{card}(A)$. Thus we have pinned down the cardinality of A to exactly λ.

It follows that Γ has a model of cardinality λ, provided that Γ' has a model. This is where the Compactness Theorem (Theorem 7.6.8) comes into play. To show that Γ' has a model, we need only show that every finite subset of it has a model. We can show this by proving that for every finite subset X of W, say with n elements, there is a model of Γ together with the sentences $(c \neq d)$ where $c, d \in X$.

If Γ has a model B with infinite cardinality, then we can choose n distinct elements $b_1, \ldots, b_n$ of B and take the witnesses in X to name these elements. This proves (a).

If there are infinitely many natural numbers m in the spectrum of Γ, then choose m at least as great as n, take B to be a model of Γ with m elements, and again choose n distinct elements of B to be named by the witnesses. This proves (b). ☐

Anatoliĭ Mal'tsev Russia, 1909–1967. A pioneer in applications of logic to algebra, he was the first to publish a construction of models with uncountable cardinality.

It follows at once from Theorem 7.9.4, and the fact that there is no greatest cardinal, that if a first-order theory has an infinite model then it has models with different cardinalities. But two models with different cardinalities cannot be isomorphic. So if a theory Γ has infinite models, then the most we can hope for is that each model of Γ is isomorphic to all other models of Γ *of the same cardinality*.

Definition 7.9.5 Let Γ be a consistent theory.

(a) We say that Γ is *categorical* if, for all models A, B of Γ, A is isomorphic to B.

(b) Let κ be a cardinal. We say that Γ is *κ-categorical* if Γ has models of cardinality κ, and all of them are isomorphic.

Categorical first-order theories do exist; but if Γ is such a theory then there is a positive integer n such that all models of Γ have exactly n elements (as, for example, in Exercise 7.9.1). Also κ-categorical theories do exist with κ infinite. When κ is uncountable, the models of κ-categorical theories have some very interesting features that relate them to algebraic geometry. But here we confine ourselves to an example of an ω_0-categorical theory.

Recall σ_{lo}, the signature of linear orders, and let Δ be the theory consisting of the axioms for a linearly ordered set (Example 7.7.1(b)) and the following two sentences:

$$\forall x \forall y \ (x < y \rightarrow \exists z((x < z) \wedge (z < y)))$$
$$\forall x \exists y \exists z \ ((y < x) \wedge (x < z))$$

A linearly ordered set X which is a model of the first sentence has the property that between any two elements of X there is another element. Such a linearly ordered set is said to be *dense*. If X is a model of the second sentence, then X does not have a greatest or a least element (see Definition 5.2.5). Such a linearly ordered set is said to be *without endpoints*. We call Δ the *theory of dense linearly ordered sets without endpoints*. (By Example 5.2.6 a finite linearly ordered set has a least element; so linearly ordered sets without endpoints are always infinite.)

We use the usual notation for intervals in a linearly ordered set. Namely, if a, b are elements of A and $a <_A b$, we define

$$(a, b) = \{x \in A \mid a <_A x <_A b\}$$

$$(-\infty, a) = \{x \in A \mid x <_A a\}$$

$$(a, \infty) = \{x \in A \mid a <_A x\}$$

The following was proved by Cantor, though the proof we use was discovered later by Edward Huntington and Felix Hausdorff.

Theorem 7.9.6 *The theory of dense linearly ordered sets without endpoints is ω_0-categorical.*

Proof In order to avoid confusion between intervals and ordered pairs, in this proof we will write ordered pairs as $\langle a, b \rangle$. Let X, Y be countable dense linearly ordered sets without endpoints. We can list the elements of X and Y without repetition, say $X = \{x_0, x_1, \dots\}$ and $Y = \{y_0, y_1, \dots\}$. We will define recursively a sequence of pairs

$$\langle a_0, b_0 \rangle, \langle a_1, b_1 \rangle, \dots$$

where each a_n is in X and each b_n is in Y, such that

(7.49)   for all m and n, $a_m <_X a_n$ if and only if $b_m <_Y b_n$.

We begin by choosing $\langle a_0, b_0 \rangle = \langle x_0, y_0 \rangle$.

Then for each even number $n > 0$, assuming the pairs up to $\langle a_{n-1}, b_{n-1} \rangle$ have been chosen so that (7.49) holds, we choose a_n to be the first element in the listing of X that does not appear among $a_0, \dots, a_{n-1}$. Now the set

$\{a_0, \ldots, a_{n-1}\}$ is linearly ordered as a subset of X, say as $x_{i_0} <_X x_{i_1} <_X \cdots <_X$ $x_{i_{m-1}}$ for some $m \leqslant n$ (see Example 5.2.6). This divides X into $m + 1$ intervals: the intervals $(x_{i_j}, x_{i_{j+1}})$ for $0 \leq j < m - 1$, and the two intervals $(-\infty, x_{i_0})$ and $(x_{i_{m-1}}, \infty)$. The assumption (7.49) implies that the elements $b_0, \ldots, b_{n-1}$ divide Y into $m + 1$ intervals corresponding to the division of X. We chose a_n to be distinct from $a_0, \ldots, a_{n-1}$, so it lies in one of the intervals of X. Suppose this interval is (a_j, a_k). Then $a_j <_X a_n <_X a_k$, and we can use the denseness of Y to find b_n so that $b_j <_Y b_n <_Y b_k$. For definiteness, let b_n be the first element in the listing of Y that lies in this interval. The argument when a_n lies in one of the two end intervals is similar, using the fact that Y has no endpoints. In all cases (7.49) is true up to the pair $\langle a_n, b_n \rangle$.

For each odd number n, do the same but with X and Y the other way round, so that we first choose b_n and then find a_n to match.

Now each element x_i of X was chosen to be a_n for some $n \leqslant 2i$, and likewise each element y_i of Y was chosen to be b_n for some $n \leqslant 2i + 1$. Also if $b_m \neq b_n$ then $a_m \neq a_n$, since if $b_m <_Y b_n$ then $a_m <_X a_n$ by (7.49) (and likewise if $b_n <_Y b_m$). So we can define a surjective function $f : X \to Y$ by putting $f(a_n) = b_n$ for each $n \in \mathbb{N}$. This function f is injective since $a_m <_X a_n$ implies $b_m <_Y b_n$. By (7.49), f is an order-isomorphism from X to Y. □

Notice how the proof chose a_1 to match b_1, then b_2 to match a_2, then a_3 to match b_3 and so on. This alternation is known as *back-and-forth*, and it has become enormously influential in applications of logic, particularly in theoretical computer science.

Exercises

7.9.1. Let σ be the empty signature. Every σ-structure is just a set (its domain), so that two σ-structures are isomorphic if and only if they have the same cardinality. Show that for each positive integer n there is a sentence ψ_n of LR(σ) whose models are exactly the sets of cardinality n (so that $\{\psi_n\}$ is categorical).

7.9.2. (a) Let Γ be a consistent theory in LR(σ), where σ is countable. Assume

    (1)   For some infinite cardinal κ, Γ is κ-categorical;

    (2)   all models of Γ are infinite.

Show that the theory Γ is complete. [If not, then by the Spectrum Theorem it would have two models of cardinality κ in which different sentences are true. The clauses (1) and (2) are sometimes known as the *Loś-Vaught test for completeness*.]

(b) Deduce that if A and B are any two dense linearly ordered sets without endpoints, then $A \equiv B$. (Taking the linearly ordered sets of the rationals and the reals, which have different cardinalities, this provides a concrete counterexample to the converse of Theorem 7.9.2.)

7.9.3. Show that the class of all well-ordered sets is not a model class. [The proof is very similar to that of Theorem 7.9.4. For contradiction suppose the class of well-ordered sets is $\mathrm{Mod}(\Gamma)$. Introduce countably many new constant symbols $c_0, c_1, \ldots$, add to Γ the sentences

$$\ldots, \quad c_3 < c_2, \quad c_2 < c_1, \quad c_1 < c_0$$

and use the Compactness Theorem (Theorem 7.6.8).]

7.9.4. Prove that every countable linearly ordered set X is order-isomorphic to a subset Z of $\mathbb{Q}$, where Z is ordered with the restriction of the usual linear order on $\mathbb{Q}$. [Use a simplified version of the argument for Theorem 7.9.6, with $Y = \mathbb{Q}$. You only need to go forth, not back.]

7.9.5. Show that if $f : A \to B$ is an isomorphism of σ-structures then $f^{-1} : B \to A$ exists and is also an isomorphism. If moreover $g : B \to C$ is an isomorphism of σ-structures, show that the composed function $g \circ f : A \to C$ is an isomorphism.

7.9.6. As in Exercise 5.5.7, an *automorphism* of a σ-structure A is an isomorphism from A to A. The identity function on the domain of a σ-structure A is always an automorphism of A; we say A is *rigid* if this is the only automorphism of A.

(a) Show that every finite linearly ordered set is rigid.

(b) Show that the linearly ordered set of the natural numbers is rigid.

(c) Show that if a linearly ordered set A is not rigid then it has infinitely many automorphisms. [If $f(a) < a$ then $f(f(a)) < f(a)$, so f and $f \circ f$ are distinct automorphisms.]

(d) Show that the linearly ordered set of the rationals has uncountably many automorphisms.

8   Postlude

We say that a sentence of LR(σ) is *valid* if it is true in every σ-structure. We will prove in a moment that, unlike LP, there is no algorithm for telling whether a given sentence of LR is valid. Nevertheless we have the following. Recall (Definition 7.7.5(a)) that ϕ is a *consequence* of Γ if $\Gamma \models \phi$.

Theorem 8.1 *For any finite signature σ, let Γ be a computably enumerable (c.e.) set of sentences of LR(σ). Then the set of consequences of Γ is c.e.*

Proof Since Γ is c.e., we can set up a computer $C1$ to list the sentences in Γ. (When Γ is empty the list is empty.)

We claim that we can also set up a computer $C2$ to list all and only the σ-derivations. In Section 3.4 we saw that every σ derivation is a labelled tree, and we saw in Exercise 3.2.5 how to give a labelled tree a Gödel number. So we can devise a system giving each σ-derivation D a Gödel number GN(D); as with formulas, each of D and GN(D) can be calculated from the other. The computer $C2$ goes through the natural numbers, and at each n it checks whether n is GN(D) for some σ-derivation D whose conclusion is a sentence; it can do this by Theorem 7.4.4. When the answer is Yes, it prints out the derivation D.

Now to prove the theorem, set up a third computer $C3$ which does the following for each natural number n in turn. It looks at each of the first n derivations D listed by $C2$, and for each D it checks whether the undischarged assumptions of D are all among the first n sentences listed by $C1$. If Yes, then $C3$ prints out the conclusion of D. Thus $C3$ will enumerate all and only the sentences ϕ such that $\Gamma \vdash_\sigma \psi$. By the Completeness Theorem (Theorem 7.6.7) these sentences are precisely the consequences of Γ.   □

Theorem 8.2 (Gödel's Diagonalisation Theorem) *There is no computable set of sentences of LR(σ_{arith}) true in $\mathbb{N}$ that includes all the true diophantine sentences.*

Proof We will use the Gödel numbers of sentences in place of the sentences themselves. This is allowable since each is computable from the other. Suppose for contradiction that X is the set of Gödel numbers of a set of sentences true in $\mathbb{N}$ that includes all the true diophantine sentences, and that X is computable. Then the set $Y = \mathbb{N} \backslash X$ is c.e., by Lemma 5.8.5(a) $\Rightarrow$ (b).

Classification of sentences about ℕ:

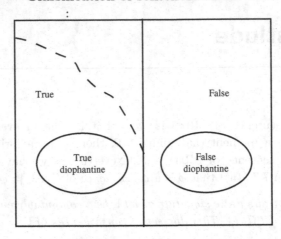

For any m and n, let $S(m,n)$ be the statement

m is the Gödel number of a formula θ of $\mathrm{LR}(\sigma_{\mathrm{arith}})$, and Y
contains the Gödel number of the formula $\theta[\bar{n}/x]$.

Since Y is c.e., we can check that the set of pairs (m,n) such that $S(m,n)$
holds is also c.e.

So by Matiyasevich's Theorem (Theorem 5.8.6) there is a diophantine formula $\chi(x,y)$ which expresses $S(x,y)$. We can assume by Example 7.3.8 that no
quantifier $\forall x$ or $\exists x$ (with the variable x) occurs in χ, so that x is substitutable for
y in χ. Let n be the Gödel number of the formula $\chi[x/y]$. Then $\chi[\bar{n}/x,\bar{n}/y]$ is true
in ℕ if and only if Y contains the Gödel number of the sentence $\chi[\bar{n}/x,\bar{n}/y]$. So

$$\mathrm{GN}(\chi[\bar{n}/x,\bar{n}/y]) \in X \quad \Leftrightarrow \quad \models_{\mathbb{N}} \chi[\bar{n}/x,\bar{n}/y] \qquad \text{(since if } \theta \text{ is diophan-}$$
$$\text{tine, } X \text{ contains } \mathrm{GN}(\theta)$$
$$\text{if and only if } \theta \text{ is true)}$$
$$\Leftrightarrow \quad \mathrm{GN}(\chi[\bar{n}/x,\bar{n}/y]) \in Y \qquad \text{(by choice of } \chi \text{ and } n\text{)}$$
$$\Leftrightarrow \quad \mathrm{GN}(\chi[\bar{n}/x,\bar{n}/y])\rceil \notin X \qquad \text{(since } X \text{ is the comple-}$$
$$\text{ment of } Y\text{)}$$

contradiction. □

Corollary 8.3 (Matiyasevich's solution of Hilbert's Tenth Problem) There
is no algorithm to determine, given an equation of $\mathrm{LR}(\sigma_{\mathrm{arith}})$, whether the equation has a solution in ℕ. (Hilbert's Tenth Problem (1900) was: Either find such
an algorithm or show that none exists.)

Kurt Gödel Austria and USA, 1906–1978.
The set of first-order sentences true in $\mathbb{N}$ is not computably enumerable.

Proof Let Γ be the set of all diophantine sentences that are true in $\mathbb{N}$. By Theorem 8.2, Γ is not computable. But if there was an algorithm as described in the Corollary, we could use it to compute whether or not a sentence is in Γ. So there is no such algorithm. $\square$

Corollary 8.4 (Church's Undecidability Theorem) There is no algorithm to determine, given a sentence ϕ of first-order logic, whether or not ϕ is valid.

Proof Let θ be the conjunction of the four sentences of PA_0 (Definition 7.5.2), and let ψ be a diophantine sentence. Then the following are equivalent:

- ψ is true in $\mathbb{N}$.
- $\{\mathrm{PA}_0\} \vdash \psi$ (by Dedekind's Theorem, Theorem 7.5.3)
- $\vdash (\theta \to \psi)$ ($\Rightarrow$ by ($\wedge$E) and ($\to$I), $\Leftarrow$ by ($\to$E) and ($\wedge$I))
- $\models (\theta \to \psi)$ (by the Completeness Theorem, Theorem 7.6.7).

So an algorithm for deciding whether or not a sentence of $\mathrm{LR}(\sigma_{\mathrm{arith}})$ is valid would give us an algorithm for deciding whether or not a diophantine sentence ψ is true in $\mathbb{N}$. But by Corollary 8.3 there is no such algorithm. $\square$

Alonzo Church USA, 1903–1995.
Computability is mathematically definable.

Corollary 8.5 (Gödel's Undecidability Theorem) For every c.e. set Γ of sentences of $\mathrm{LR}(\sigma_{\mathrm{arith}})$ that are true in $\mathbb{N}$, there is some sentence of $\mathrm{LR}(\sigma_{\mathrm{arith}})$ that is true in $\mathbb{N}$ but is not in Γ.

Proof Suppose for contradiction that Γ is exactly the set of sentences of $\mathrm{LR}(\sigma_{\mathrm{arith}})$ true in $\mathbb{N}$. Then the set Δ of sentences false in $\mathbb{N}$ is $\{\phi \mid \neg\phi \in \Gamma\}$, which is computably enumerable using an enumeration of Γ. So using enumerations of Γ and Δ, the set Γ is computable by Lemma 5.8.5(b) $\Rightarrow$ (a). But Theorem 8.2 tells us that Γ is not computable, since it is a set of true sentences and it contains all true diophantine sentences. $\square$

In Example 7.7.3 we described a set PA of sentences that are true in $\mathbb{N}$.

Corollary 8.6 (Gödel's Incompleteness Theorem) There are sentences of $\mathrm{LR}(\sigma_{\mathrm{arith}})$ that are true in $\mathbb{N}$ but are not consequences of PA.

Proof The set PA is a computable set of sentences that are true in $\mathbb{N}$. So by Theorem 8.1 the set of consequences of PA is a c.e. set of sentences true in $\mathbb{N}$. Now apply Corollary 8.5. $\square$

Gödel's Undecidability Theorem saves us from the awful possibility that someone might write a computer program that solves all problems of mathematics. Even for arithmetic no such program can exist.

Appendix A   The natural deduction rules

NATURAL DEDUCTION RULE (AXIOM RULE)   Let ϕ be a statement. Then

$$\phi$$

is a derivation. Its conclusion is ϕ, and it has one undischarged assumption, namely ϕ.

NATURAL DEDUCTION RULE ($\wedge$I)   If

$$\begin{array}{ccc} D & & D' \\ \phi & \text{and} & \psi \end{array}$$

are derivations of ϕ and ψ, respectively, then

$$\frac{\begin{array}{cc} D & D' \\ \phi & \psi \end{array}}{(\phi \wedge \psi)} \ (\wedge\text{I})$$

is a derivation of $(\phi \wedge \psi)$. Its undischarged assumptions are those of D together with those of D'.

NATURAL DEDUCTION RULE ($\wedge$E)   If

$$\begin{array}{c} D \\ (\phi \wedge \psi) \end{array}$$

is a derivation of $(\phi \wedge \psi)$, then

$$\frac{\begin{array}{c} D \\ (\phi \wedge \psi) \end{array}}{\phi} \ (\wedge\text{E}) \quad \text{and} \quad \frac{\begin{array}{c} D \\ (\phi \wedge \psi) \end{array}}{\psi} \ (\wedge\text{E})$$

are derivations of ϕ and ψ, respectively. Their undischarged assumptions are those of D.

Natural Deduction Rule ($\to$I)   Suppose

$$
\begin{array}{c}
D \\
\psi
\end{array}
$$

is a derivation of ψ, and ϕ is a formula. Then the following is a derivation of $(\phi \to \psi)$:

$$
\frac{\begin{array}{c} \not\phi \\ D \\ \psi \end{array}}{(\phi \to \psi)} \ (\to\text{I})
$$

Its assumptions are those of D, except possibly ϕ.

Natural Deduction Rule ($\to$E)   If

$$
\begin{array}{cc}
D & \quad D' \\
\phi & \text{and} \quad (\phi \to \psi)
\end{array}
$$

are derivations of ϕ and $(\phi \to \psi)$, respectively, then

$$
\frac{\begin{array}{cc} D & \quad D' \\ \phi & \quad (\phi \to \psi) \end{array}}{\psi} \ (\to\text{E})
$$

is a derivation of ψ. Its assumptions are those of D together with those of D'.

Natural Deduction Rule ($\leftrightarrow$I)   If

$$
\begin{array}{cc}
D & \quad D' \\
(\phi \to \psi) & \text{and} \quad (\psi \to \phi)
\end{array}
$$

are derivations of $(\phi \to \psi)$ and $(\psi \to \phi)$, respectively, then

$$
\frac{\begin{array}{cc} D & \quad D' \\ (\phi \to \psi) & \quad (\psi \to \phi) \end{array}}{(\phi \leftrightarrow \psi)} \ (\leftrightarrow\text{I})
$$

is a derivation of $(\phi \leftrightarrow \psi)$. Its undischarged assumptions are those of D together with those of D'.

NATURAL DEDUCTION RULE ($\leftrightarrow$E)   If

$$D$$
$$(\phi \leftrightarrow \psi)$$

is a derivation of $(\phi \leftrightarrow \psi)$, then

$$\frac{\begin{array}{c} D \\ (\phi \leftrightarrow \psi) \end{array}}{(\phi \rightarrow \psi)}\ (\leftrightarrow\text{E}) \quad \text{and} \quad \frac{\begin{array}{c} D' \\ (\phi \leftrightarrow \psi) \end{array}}{(\psi \rightarrow \phi)}\ (\leftrightarrow\text{E})$$

are derivations of $(\phi \rightarrow \psi)$ and $(\psi \rightarrow \phi)$, respectively. Their undischarged assumptions are those of D.

NATURAL DEDUCTION RULE ($\neg$E)   If

$$\begin{array}{c} D \\ \phi \end{array} \quad \text{and} \quad \begin{array}{c} D' \\ (\neg\phi) \end{array}$$

are derivations of ϕ and $(\neg\phi)$, respectively, then

$$\frac{\begin{array}{cc} D & D' \\ \phi & (\neg\phi) \end{array}}{\bot}\ (\neg\text{E})$$

is a derivation of $\bot$. Its undischarged assumptions are those of D together with those of D'.

NATURAL DEDUCTION RULE ($\neg$I)   Suppose

$$D$$
$$\bot$$

is a derivation of $\bot$, and ϕ is a statement. Then the following is a derivation of $\neg\phi$):

$$\frac{\begin{array}{c} \cancel{\phi} \\ D \\ \bot \end{array}}{(\neg\phi)}\ (\neg\text{I})$$

Its undischarged assumptions are those of D, except possibly ϕ.

NATURAL DEDUCTION RULE (RAA)   Suppose we have a derivation

$$D$$
$$\bot$$

whose conclusion is $\perp$. Then there is a derivation

$$
\begin{array}{c}
\cancel{(\neg\phi)} \\
D \\
\hline
\perp \\
\phi
\end{array} \quad \text{(RAA)}
$$

Its assumptions are those of D, except possibly $(\neg\phi)$.

NATURAL DEDUCTION RULE ($\vee$I)   If

$$
\begin{array}{c}
D \\
\phi
\end{array}
$$

is a derivation with conclusion ϕ, then

$$
\begin{array}{c}
D \\
\phi \\
\hline
(\phi \vee \psi)
\end{array}
$$

is a derivation of $(\phi\vee\psi)$. Its undischarged assumptions are those of D. Similarly if

$$
\begin{array}{c}
D \\
\psi
\end{array}
$$

is a derivation with conclusion ψ, then

$$
\begin{array}{c}
D \\
\psi \\
\hline
(\phi \vee \psi)
\end{array}
$$

is a derivation with conclusion $(\phi \vee \psi)$. Its undischarged assumptions are those of D.

NATURAL DEDUCTION RULE ($\vee$E)   Given derivations

$$
\begin{array}{c}
D \\
(\phi \vee \psi)
\end{array} , \quad
\begin{array}{c}
D' \\
\chi
\end{array} \quad \text{and} \quad
\begin{array}{c}
D'' \\
\chi
\end{array}
$$

we have a derivation

$$
\begin{array}{ccc}
 & \cancel{\phi} & \cancel{\psi} \\
D & D' & D'' \\
(\phi \vee \psi) & \chi & \chi \\
\hline
 & \chi &
\end{array}
$$

Its undischarged assumptions are those of D, those of D' except possibly ϕ, and those of D'' except possibly ψ.

NATURAL DEDUCTION RULE (=I)   If t is a term then

$$\frac{}{(t=t)}\ (\text{=I})$$

is a derivation of the formula $(t = t)$. It has no assumptions.

NATURAL DEDUCTION RULE (=E)   If ϕ is a formula, s and t are terms substitutable for x in ϕ, and

$$\begin{array}{cc} D & D' \\ (s = t) & \phi[s/x] \end{array}\ ,$$

are derivations, then so is

$$\frac{\begin{array}{cc} \begin{array}{c} D \\ (s = t) \end{array} & \begin{array}{c} D' \\ \phi[s/x] \end{array} \end{array}}{\phi[t/x]}\ (\text{=E})$$

Its undischarged assumptions are those of D together with those of D'.

NATURAL DEDUCTION RULE ($\forall$E)   Given a derivation

$$\begin{array}{c} D \\ \forall x \phi \end{array}$$

and a term t that is substitutable for x in ϕ, the following is also a derivation:

$$\frac{\begin{array}{c} D \\ \forall x \phi \end{array}}{\phi[t/x]}$$

Its undischarged assumptions are those of D.

NATURAL DEDUCTION RULE ($\exists$I)   Given a derivation

$$\begin{array}{c} D \\ \phi[t/x] \end{array}$$

where t is a term that is substitutable for x in ϕ, the following is also a derivation:

$$\frac{\begin{array}{c} D \\ \phi[t/x] \end{array}}{\exists x \phi}$$

Its undischarged assumptions are those of D.

NATURAL DEDUCTION RULE ($\forall$I)   If

$$D$$
$$\phi[t/x]$$

is a derivation, t is a constant symbol or a variable, and t does not occur in ϕ or in any undischarged assumption of D, then

$$D$$
$$\frac{\phi[t/x]}{\forall x \phi}$$

is also a derivation. Its undischarged assumptions are those of D.

NATURAL DEDUCTION RULE ($\exists$E)   If

$$\begin{array}{ccc} D & \text{and} & D' \\ \exists x \phi & & \chi \end{array}$$

are derivations and t is a constant symbol or a variable which does not occur in χ, ϕ or any undischarged assumption of D' except $\phi[t/x]$, then

$$\begin{array}{cc} & \phi[t/x] \\ D & D' \\ \dfrac{\exists x \phi \qquad\qquad \chi}{\chi} \end{array}$$

is also a derivation. Its undischarged assumptions are those of D' except possibly $\phi[t/x]$, together with those of D.

Appendix B   Denotational semantics

It is a very old idea that sentences are built up by two parallel processes: combining words on the page, and combining meanings in the mind. Already in the tenth century the Arabic philosopher Al-Fārābī talked of 'the imitation of the composition of meanings by the composition of expressions'. As long as people agreed with Al-Fārābī that the meanings wear the trousers, there was no way in for mathematics, because nobody had any idea where to look for the mathematical structure of meanings.

During the period 1850–1930 the idea gradually took root that at least for some artificial languages, we can describe the syntax independently of meanings, and then we can describe how meanings of complex phrases are built up from the meanings of words, using the syntax as a template. In 1933 Tarski showed exactly how to build up a semantics in this way for most formal languages of logic. In 1963 Helena Rasiowa and Roman Sikorski popularised an algebraic version of Tarski's theory: formal languages are algebras with the rules of syntactic composition as their operations, and we interpret a language by describing a homomorphism from the algebra to a suitable structure (e.g. a boolean algebra).

Around 1970 these ideas spread into two new areas. First, Dana Scott and Christopher Strachey showed how to extend Tarski's framework to computer languages. Second, Richard Montague and Barbara Partee launched a programme to carry the same ideas over into natural languages. Scott and Strachey advertised their scheme as 'denotational semantics', while Partee used

Helena Rasiowa Poland, 1917–1994.
We can handle logic by methods of algebra.

the catchword 'compositionality', but the ideas involved had a good deal in common. (Both Scott and Montague had worked closely with Tarski.) The work of these four people has been hugely influential in computer science and in linguistics.

Like many people today, in this book we have taken the view that sentences have an inner framework; we represent it by their parsing trees. We *interpret* sentences by climbing up their parsing trees, and we also work out how to *write or speak* them by climbing up their parsing trees. Somebody else can worry about whether parsing trees themselves are really syntactic or semantic.

Our main aim in this appendix is to show how the tree-climbing analyses of this book are related to more algebraic accounts of semantics.

Recall the truth table that described the behaviour of truth function symbols. Here we write it using 1 for truth and 0 for falsehood:

	ϕ	ψ	$(\phi \wedge \psi)$	$(\phi \vee \psi)$	$(\phi \to \psi)$	$(\phi \leftrightarrow \psi)$	$(\neg\phi)$	$\bot$
	1	1	1	1	1	1	0	0
(B.1)	1	0	0	1	0	0		
	0	1	0	1	1	0	1	
	0	0	0	0	1	1		

One can also read this table as a set of definitions of functions:

	ϕ	ψ	$b_\wedge$	$b_\vee$	$b_\to$	$b_\leftrightarrow$
	1	1	1	1	1	1
(B.2)	1	0	0	1	0	0
	0	1	0	1	1	0
	0	0	0	0	1	1

Here $b_\wedge$ is a function which takes two truth values to a truth value; for example, $b_\wedge(1,1) = 1$ and $b_\wedge(1,0) = 0$ according to the table. Also $b_\neg$ is a function of one truth value, with $b_\neg(1) = 0$ and $b_\neg(0) = 1$; and $b_\bot$ is the constant function with value 0. The b stands for *boolean function*.

In Section 3.5 we saw how to use a σ-structure A to assign a truth value to a propositional formula ϕ. The structure tells us what truth values to write on the leaves, and the truth tables tell us how to put truth values as we climb up the parsing tree of ϕ. Here are the rules for the assignment, written as a compositional definition.

Let σ be a signature and A a σ-structure. We write $A^\star$ for the following compositional definition:

(B.3)
$$A(\chi) \circ \chi \qquad 0 \circ \perp$$

```
   b_¬(v) o ¬         b_□(v,w) o □
         |                /   \
       v o            v o      o w
```

where χ is a propositional symbol and $\square \in \{\wedge, \vee, \rightarrow, \leftrightarrow\}$.

If ϕ is a formula with parsing tree π, we define $A^\star(\phi)$, the *truth value of ϕ at A*, to be $A^\star(\pi)$.

Linguists and computer scientists would comment at this point that we are only halfway done. We have said what $A^\star(\chi)$ is; but A is not part of the language. We need to assign to χ a *semantic value*, in symbols $|\chi|$, that tells us what the truth value of χ is for each possible structure A. In other words, $|\chi|$ should give us $A^\star(\chi)$ *as a function of A*.

The following notation (called *lambda-notation*) is standard for this purpose. We write $\lambda x(x - y)$ for $x - y$ *as a function of x*. So, for example,

$$(\lambda x(x - y))(6) = 6 - y.$$

Taking the same idea a step further,

$$(\lambda y \lambda x(x - y))(6)(4) = \lambda x(x - 4)(6) = 6 - 4 = 2$$

whereas $(\lambda x \lambda y(x - y))(6)(4)$ works out as -2. If C is a σ-structure then

$$(\lambda A(A(p_0)))(C) = C(p_0)$$

so that $\lambda A(A(p_0))$ is the function that takes each σ-structure A to the value that A gives to p_0.

In this notation we have the following *denotational semantics for propositional logic*. We assume a signature σ, and A ranges over all possible σ-structures.

$$\lambda A(A(\chi)) \circ \chi \qquad 0 \circ \perp$$

```
  λA(b_¬(v(A)))  o ¬      λA(b_□(v(A),w(A)))  o □
         |                        /     \
       v o                    v o         o w
```

(B.4)

Applying this compositional definition to the parsing tree of χ, we get $|\chi|$ as the label on the root.

We turn to first order logic. Just as with boolean functions, we need a more functional notation than we used in the book. Given a structure A, we define an *assignment in A* to be a function α whose domain is a set of variables, which takes each variable to an element of A. If E is an expression whose free variables are all in the domain of α, then we can regard α as a set of instructions for substituting a name $\overline{\alpha(v)}$ for each free variable v in E. If E is a term, this substitution leads to an element of A, whereas if E is a formula, it leads to a truth value.

We first interpret terms, then formulas. The interpretation of a term t of LR(σ) depends on a σ-structure A and an assignment α in A whose domain includes $FV(t)$. Treating A as fixed, we read off the interpretation of t in A from the following compositional definition:

$$
\text{(B.5)} \quad \lambda\alpha(\alpha(v)) \circ v \qquad \lambda\alpha(c_A) \circ c \qquad\qquad \lambda\alpha(F_A(a_1(\alpha),\dots,a_n(\alpha)))
$$

where v is a variable and F is a function symbol of arity n.

Applying this definition to a term t gets us a function t_A which, applied to an assignment α whose domain includes every variable in t, gives the element of A named by t under this assignment. You can adapt the definition to get $|t| = \lambda A(t_A)$ as a function of the structure A. You might also consider what is needed to make t_A into a function defined only on the assignments whose domain is *exactly* $FV(t)$; but every piece of tidying adds a new layer of clutter to the definition.

Next we consider relation symbols. If R is an n-ary relation symbol on a set X, its *characteristic function* is the function $\chi_R : X^n \to \{1,0\}$ defined by:

$$
\chi_R(a_1,\dots,a_n) = \begin{cases} 1 & \text{if } (a_1,\dots,a_n) \in R \\ 0 & \text{otherwise} \end{cases}
$$

Likewise, we define $\chi_=$, the characteristic function of equality, to be the function from X^2 to $\{1,0\}$ such that

$$
\chi_=(a_1,a_2) = \begin{cases} 1 & \text{if } a_1 = a_2 \\ 0 & \text{otherwise} \end{cases}
$$

At atomic formulas the denotational semantics has the rules

$$
\lambda\alpha0 \circ \bot \qquad \lambda\alpha(\chi_=(a_1(\alpha),a_2(\alpha))) \qquad \lambda\alpha(\chi_R(a_1(\alpha),\dots,a_n(\alpha)))
$$

$$
\text{(B.6)}
$$

where R is a relation symbol of arity n.

So again the left labels are functions f defined on assignments; but now the values $f(\alpha)$ are truth values (1 or 0).

The definition extends to truth functions by adapting (B.3) to take into account the assignments:

(B.7)

$$\lambda\alpha b_\neg(f(\alpha)) \quad \neg \qquad\qquad \lambda\alpha b_\square(f(\alpha), g(\alpha)) \quad \square$$
$$f \qquad\qquad\qquad\qquad\qquad f \qquad\qquad g$$

where $\square \in \{\wedge, \vee, \rightarrow, \leftrightarrow\}$.

It remains only to add clauses for the quantifiers. We write '$a \in A$' as shorthand for 'a is an element of A'. We write $\alpha(a/x)$ for the assignment β defined by

$$\beta(y) = \begin{cases} a & \text{if } y \text{ is } x \\ \alpha(y) & \text{if } \alpha(y) \text{ is defined and } y \text{ is not } x \end{cases}$$

If X is a non-empty subset of $\{0, 1\}$ then $\min X$ and $\max X$ are the minimum and the maximum element of X.

(B.8)

$$\lambda\alpha \min\{f(\alpha(a/x)) \mid a \in A\} \quad \forall x \qquad\qquad \lambda\alpha \max\{f(\alpha(a/x)) \mid a \in A\} \quad \exists x$$
$$f \qquad\qquad\qquad\qquad\qquad\qquad\qquad f$$

where x is a variable.

When we add λA at the front of each left label as before, we create a function that has all σ-structures in its domain, so that its domain is a proper class. This move is legitimate, but it does create some set-theoretic complications. A course in set theory is the place to discuss them.

Appendix C   Solutions to some exercises

2.1.1. (Possible answer)
        Conclusion: r has a square root.
        Assumptions:
        - r is a positive real number.
        - $x^2 - r$ is a continuous function on $\mathbb{R}$.
        - The Intermediate Value Theorem.
        - Various other arithmetical assumptions that are not spelt out.

2.1.2. (Possible answer)
        Conclusion: $\lim_{x \to 0} u(x) = 1$.
        Assumptions:
        - $1 - \frac{x^2}{4} \leqslant u(x) \leqslant 1 + \frac{x^2}{2}$ for all $x \neq 0$.
        - $\lim_{x \to 0}(1 - (x^2/4)) = 1$.
        - $\lim_{x \to 0}(1 + (x^2/2)) = 1$.
        - The Sandwich Theorem.
        - Various arithmetical assumptions.

2.1.3. **Rule A:** Intuitively correct; for the formal version see the solution to Exercise 3.4.6(b).

        Rule B: False. For example, if we are talking about the real numbers, '$x = 1$' entails '$x > 0$', the reverse entailment does not hold.

        Rule C: False. For an example in the real numbers again, if Γ consists of the statement $x = 1$ and Δ consists of the statement $x = 2$, then each of Γ and Δ entails $x > 0$. But there is no statement that is in both Γ and Δ, so $(\Gamma \cap \Delta)$ is the empty set, which does not entail $x > 0$.

2.2.1. (a) (The real number r is positive $\wedge$ the real number r is not an integer).
        [Best to repeat 'the real number r' in the second clause, rather than saying 'it is not an integer'. The reason is that in logical manipulations we want to be able to move the statements around, and moving the second clause would break the link from 'it' to 'the real number r'.]

    (b) (v is a vector $\land$ $v \neq 0$).

    (c) (If ϕ then ψ $\land$ if ψ then ϕ).

2.3.2.

$$\frac{\dfrac{\begin{array}{c}D\\\phi\end{array} \qquad \begin{array}{c}D'\\\psi\end{array}}{(\phi \land \psi)} \quad (\land\text{I})}{\phi} \quad (\land\text{E})$$

The shorter derivation is

$$\begin{array}{c}D\\\phi\end{array}$$

2.4.1.   (a) (f is differentiable $\to$ f is continuous). [There is a slight catch here. Depending on the context, (a) could mean that the 'If ... then' is true for *all* appropriate f, as if it said

For all functions $f : \mathbb{R} \to \mathbb{R}$, f is continuous if f is differentiable.

Propositional logic does not have any rules for handling 'For all'. We will meet the appropriate rules in Section 7.4.]

    (b) (x is positive $\to$ x has a square root).

    (c) ($a \neq 0 \to ab/b = a$).

2.4.2.   (a)

$$\cfrac{\cfrac{\cfrac{(\phi \land \psi)^{①}}{\psi}\,(\land\text{E}) \qquad \cfrac{(\phi \land \psi)^{①}}{\phi}\,(\land\text{E})}{(\psi \land \phi)}\,(\land\text{I})}{((\phi \land \psi) \to (\psi \land \phi))}\,{①}\,(\to\text{I})$$

    (b)

$$\cfrac{\cfrac{\cfrac{\cfrac{\phi^{①} \qquad (\phi \to \psi)^{②}}{\psi}\,(\to\text{E}) \qquad (\psi \to \chi)^{③}}{\chi}\,(\to\text{E})}{(\phi \to \chi)}\,{①}\,(\to\text{I})}{((\phi \to \psi) \to (\phi \to \chi))}\,{②}\,(\to\text{I})}{((\psi \to \chi) \to ((\phi \to \psi) \to (\phi \to \chi)))}\,{③}\,(\to\text{I})$$

2.4.6.   (a) $\Rightarrow$ (b): Suppose D_1 is a derivation of ψ whose undischarged assumptions are all in $\Gamma \cup \{\phi\}$. Write $D^\star$ for a copy of D_1 in which each occurrence of the assumption ϕ is replaced by

Then take D_1' to be the derivation

$$\textcircled{1}\ \frac{\begin{array}{c} D^\star \\ \psi \end{array}}{(\phi \rightarrow \psi)}\ (\rightarrow\!I)$$

2.5.1.   (b) $\vdash (\phi \leftrightarrow \phi)$

$$\frac{\textcircled{1}\dfrac{\cancel{\phi}^{\textcircled{1}}}{(\phi \rightarrow \phi)}\ (\rightarrow\!I) \qquad \textcircled{2}\dfrac{\cancel{\phi}^{\textcircled{2}}}{(\phi \rightarrow \phi)}\ (\rightarrow\!I)}{(\phi \leftrightarrow \phi)}\ (\leftrightarrow\!I)$$

2.5.3.   Let D be a derivation of ψ with no undischarged assumptions. Then the following is a proof of $((\phi \leftrightarrow \psi) \leftrightarrow \phi)$ with no undischarged assumptions.

$$\textcircled{1}\frac{\dfrac{\begin{array}{c} D \\ \psi \end{array} \quad \dfrac{\cancel{(\phi \leftrightarrow \psi)}^{\textcircled{1}}}{(\psi \rightarrow \phi)}\ (\leftrightarrow\!E)}{\phi}\ (\rightarrow\!E)}{((\phi \leftrightarrow \psi) \rightarrow \phi)}\ (\rightarrow\!I) \qquad \textcircled{2}\frac{\dfrac{\dfrac{\begin{array}{c} D \\ \psi \end{array}}{(\phi \rightarrow \psi)}\ (\rightarrow\!I) \quad \dfrac{\cancel{\phi}^{\textcircled{2}}}{(\psi \rightarrow \phi)}\ (\rightarrow\!I)}{(\phi \leftrightarrow \psi)}\ (\leftrightarrow\!I)}{(\phi \rightarrow (\phi \leftrightarrow \psi))}\ (\rightarrow\!I)$$

$$\frac{}{((\phi \leftrightarrow \psi) \leftrightarrow \phi)}\ (\leftrightarrow\!I)$$

2.6.1.   (b) Proof of $\vdash ((\neg(\phi \rightarrow \psi)) \rightarrow (\neg\psi))$.

$$\textcircled{2}\frac{\textcircled{1}\dfrac{\dfrac{\dfrac{\cancel{\phi}^{\textcircled{1}}}{(\phi \rightarrow \psi)}\ (\rightarrow\!I) \qquad (\neg(\phi \rightarrow \psi))^{\textcircled{2}}}{\bot}\ (\neg\!E)}{(\neg\psi)}\ (\neg\!I)}{((\neg(\phi \rightarrow \psi)) \rightarrow (\neg\psi))}\ (\rightarrow\!I)$$

(c) Proof of $\vdash ((\phi \land \psi) \to (\neg(\phi \to (\neg\psi))))$.

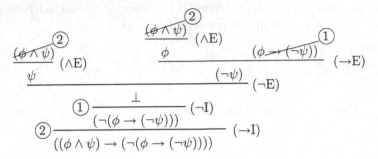

2.6.2. (b) Proof of $\vdash ((\neg(\phi \to \psi)) \to \phi)$.

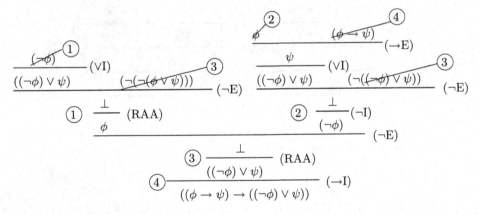

2.7.1. (c) Proof of $\vdash ((\phi \to \psi) \to ((\neg\phi) \lor \psi))$.

2.7.2. (a) Proof of $\{(\phi \lor \psi)\} \vdash (\psi \lor \phi)$.

3.1.1. (d)

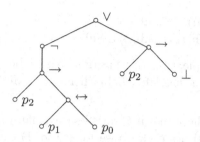

3.1.3. (f) $\{p_1, p_2, p_3, p_5, p_7\}$.

3.2.3. $\delta(\pi)$ is the length of the associated formula ϕ of π, that is, the number of symbols in ϕ.

3.2.4. The compositional definition is identical with the one in Exercise 3.2.3, except that it has 0 instead of 1 and 2 instead of 3.

3.2.5. The formula with this number is $((\neg p_1) \vee p_0)$.

3.3.4. True.

3.3.7. (c) For example, $S = \{s \mid$ the number of occurrences of '(' in s is equal to the number of occurrences of functors$\}$. Atomic formulas contain neither parentheses nor functors, so they are in S and hence (1) holds. For (2), if s is in S, then for some n it has n left parentheses and n functors, so $(\neg s)$ has $n+1$ left parentheses and $n+1$ functors and hence is also in S. The calculation for the remaining cases in (2) is similar. Hence by Exercise 3.3.6(a), S contains every formula of $\text{LP}(\sigma)$. But the expression $(p_1 \wedge p_2 \rightarrow p_3)$ has one occurrence of '(' and two functors, so it is not in S. So the expression is not a formula of $\text{LP}(\sigma)$.

(f) For example, $S = \{s \mid$ either at least one of $\wedge, \vee, \rightarrow, \leftrightarrow$ occurs in s, or there is at most one occurrence of a propositional symbol in $s\}$. For this S, (1) holds since atomic formulas contain at most one occurrence of a propositional symbol. Suppose s has at most one occurrence of a propositional symbol; then likewise so does $(\neg s)$. Also if s contains $\wedge$, $\vee$, $\rightarrow$ or $\leftarrow$, then so does $(\neg s)$. So S is closed under the first condition in (2); it is also closed under the others, since they introduce $\wedge$, $\vee$, $\rightarrow$ or $\leftrightarrow$. Hence (2) is true for S. As before, all formulas of $\text{LP}(\sigma)$ are in S. But $(\neg p_1)p_2$ is not in S, hence is not a formula of $\text{LP}(\sigma)$.

3.3.10. (b)

$$\begin{aligned}
\text{Sub}(\phi) \quad &= \{\phi\} &&\text{when } \phi \text{ is atomic,}\\
\text{Sub}((\neg\phi)) &= \text{Sub}(\phi) \cup \{\neg\phi\},\\
\text{Sub}((\phi\Box\psi)) &= \text{Sub}(\phi) \cup \text{Sub}(\psi) \cup \{(\phi\Box\psi)\} &&\text{when } \Box \in \{\wedge, \vee, \rightarrow, \leftrightarrow\}.
\end{aligned}$$

3.4.1. For example, (d)(iii): ν has right-hand label $(\wedge E)$, and for some formulas ϕ and ψ, ν has left label either ϕ or ψ and its daughter has left label $(\phi \wedge \psi)$.

3.4.3. (c) We check that if D satisfies each clause of Definition 3.4.1 for ρ then it satisfies each clause for σ too. For clause (b) it follows at once from clause (b) of this Exercise. None of the other clauses mention the signature, so their truth is independent of the signature.

(d) By Definition 3.4.4, if $(\Gamma \vdash_\rho \psi)$ is correct then there is a ρ-derivation D whose conclusion is ψ and whose undischarged assumptions all lie in Γ. By (c) of this Exercise, D is also a σ-derivation, so by Definition 3.4.4 again, $(\Gamma \vdash_\sigma \psi)$ is correct too.

3.4.4. (a) For each formula ϕ of $\text{LP}(\sigma)$, write ϕ' for the result of replacing each symbol in σ but not in ρ by $\bot$; so D' comes from D by replacing each left label ϕ by ϕ'. We have to show that D' satisfies Definition 3.4.1 for signature ρ. Clauses (a) and (c) are unaffected by passing from D to D'. Clause (b) holds since every formula ϕ' is in $\text{LP}(\rho)$. For (d), take, for example, (i); by assumption each node ν of D with right label $(\rightarrow I)$ has left label $(\phi \rightarrow \psi)$ where its daughter has left label ψ. By construction D' has left labels $(\phi \rightarrow \psi)'$ at ν and ψ' at the daughter of ν. But $(\phi \rightarrow \psi)'$ is $(\phi' \rightarrow \psi')$, so (d)(iv) is satisfied in D'. The remaining clauses of (d)–(f) are similar. Clause (g) holds for D' since it held for D and a leaf of D' carries a formula ϕ' with a dandah if and only if D carried ϕ with a dandah. So D satisfies all the clauses of Definition 3.4.1 for ρ. (You should also show that ϕ' is a formula, by induction on complexity.)

3.4.6. (b) Suppose $(\Gamma \vdash_\sigma \psi)$ is correct. Then there is a σ-derivation D whose conclusion is ψ and whose undischarged assumptions are all in Γ. Given that $\Gamma \subseteq \Delta$, the undischarged assumptions of D are also in Δ, so that D proves $(\Delta \vdash_\sigma \psi)$.

(d) By assumption there are σ-derivations D and D' proving $(\Gamma \vdash_\sigma \phi)$ and $(\Gamma \cup \{\phi\} \vdash_\sigma \psi)$, respectively. Create a new diagram D'' by taking each leaf μ of D' that carries ϕ without dandah, and attaching a copy of D so that the root of the copy takes the place of the leaf. Since the root of D has left label ϕ, this does not affect the left label at μ; but the right label of μ becomes the right label on the root of D. All the clauses (a)–(g) of Definition 3.4.1 hold for D'' since they held for D and D'; so D'' is a σ-derivation. Its conclusion is ψ. Its undischarged assumptions are those of D and D', except that the occurrences of ϕ without dandah on the leaves of D' have been removed. So the undischarged assumptions of D'' are all in Γ as required.

3.5.2. (a) Tautology and satisfiable. (b) Contradiction. (c) Satisfiable. (d) Contradiction.

3.5.4. $\Rightarrow$: Suppose $A^\star(\phi) = \mathrm{T}$ for every ρ-structure A. Let B be a σ-structure, and let A be the ρ-structure got from B by ignoring the assignments made by B to symbols not in ρ. By assumption $A^\star(\phi) = \mathrm{T}$, so $B^\star(\phi) = \mathrm{T}$ by the Principle of Irrelevance. $\Leftarrow$: Similar, but instead of ignoring the values for symbols in σ, we add them in an arbitrary way.

3.5.5. (a) 2^k.

(b) $2^k(k+n)$.

(c) For every ℓ, $\beta(\ell) \leqslant 2^{(\ell+3)/4}\left(\frac{3\ell+5}{4}\right)$; the maximum is achieved when $\ell \equiv 1 \pmod 4$.

3.6.2. Partial solution:

(a)

p	q	$(p$	$\wedge$	$q)$	$(\neg$	$(p$	$\to$	$(\neg$	$q)))$
T	T	T	T	T	T	T	F	F	T
T	F	T	F	F	F	T	T	T	F
F	T	F	F	T	F	F	T	F	T
F	F	F	F	F	F	F	T	T	F
			$\Uparrow$			$\Uparrow$			

(b)

p	q	$(p$	$\vee$	$q)$	$((p$	$\to$	$q)$	$\to$	$q)$
T	T	T	T	T	T	T	T	T	T
T	F	T	T	F	T	F	F	T	F
F	T	F	T	T	F	T	T	T	T
F	F	F	F	F	F	T	F	F	F
			$\Uparrow$					$\Uparrow$	

(c)

p	q	$(p$	$\leftrightarrow$	$q)$	$((p$	$\to$	$q)$	$\wedge$	$(q$	$\to$	$p))$
T	T	T	T	T	T	T	T	T	T	T	T
T	F	T	F	F	T	F	F	F	F	T	T
F	T	F	F	T	F	T	T	F	T	F	F
F	F	F	T	F	F	T	F	T	F	T	F
			$\Uparrow$					$\Uparrow$			

3.6.5. (a) $\Rightarrow$ (c): both ϕ and $(\neg\bot)$ are true in all σ-structures. (c) $\Rightarrow$ (d): $(\neg\bot)$ is clearly a tautology. (d) $\Rightarrow$ (a): Assume ϕ is logically equivalent to tautology ψ in $\mathrm{LP}(\sigma)$. For any σ-structure A, $A^\star(\phi) = A^\star(\psi) = \mathrm{T}$. So ϕ is a tautology.

3.7.2. (c) (i) is equivalent to (ii) by applying to the first De Morgan Law
(Example 3.6.5) the substitution

$$(p_1 \wedge (\neg p_2))/p_1, (((\neg p_1) \wedge p_2) \wedge p_3)/p_2$$

and then quoting the Substitution Theorem. To show the equivalence
of (ii) and (iii), first use De Morgan and Substitution to infer that
$(\neg(p_1 \wedge (\neg p_2)))$ is logically equivalent to $((\neg p_1) \vee (\neg(\neg p_2)))$, and then
use Replacement to put the latter in place of the former inside (ii). Next
do the same with the subformula $(\neg(((\neg p_1) \wedge p_2) \wedge p_3))$, but using (b)
above with $n = 3$ in place of De Morgan. To show the equivalence of
(iii) and (iv), use Substitution to infer from the Double Negation Law
(Example 3.6.5) that $(\neg(\neg p_2))$ eq p_2, and then use Replacement to make
this replacement in (iii); likewise Replacement allows us to write p_1 in
place of $(\neg(\neg p_1))$.

3.8.1. (b) The proof of Post's Theorem finds the DNF

$$(p_1 \wedge p_2 \wedge p_3) \vee (p_1 \wedge \neg p_2 \wedge \neg p_3) \vee$$
$$(\neg p_1 \wedge \neg p_2 \wedge p_3) \vee (\neg p_1 \wedge \neg p_2 \wedge \neg p_3)$$

The DNF of the negation of (b) is

$$(p_1 \wedge p_2 \wedge \neg p_3) \vee (p_1 \wedge \neg p_2 \wedge p_3) \vee$$
$$(\neg p_1 \wedge p_2 \wedge p_3) \vee (\neg p_1 \wedge p_2 \wedge \neg p_3)$$

Putting $\neg$ at the beginning and moving it inwards by De Morgan gives
the CNF

$$(\neg p_1 \vee \neg p_2 \vee p_3) \wedge (\neg p_1 \vee p_2 \vee \neg p_3) \wedge$$
$$(p_1 \vee \neg p_2 \vee \neg p_3) \wedge (p_1 \vee \neg p_2 \vee p_3)$$

3.8.2. (b) $(p_1 \wedge \neg p_2 \wedge \neg p_3) \vee (p_1 \wedge p_2 \wedge p_3 \wedge \neg p_4)$. You cannot cancel $\neg p_2 \wedge$
$\neg p_3$ against $p_2 \wedge p_3$, because they involve more than one propositional
symbol.

3.8.6. Start by noting that $(\phi|\phi)$ has the same truth table as $\neg\phi$.

		$\rightarrow$
1	1	1
1	$\frac{1}{2}$	$\frac{1}{2}$
1	0	0
$\frac{1}{2}$	1	1
$\frac{1}{2}$	$\frac{1}{2}$	1
$\frac{1}{2}$	0	0
0	1	1
0	$\frac{1}{2}$	1
0	0	1

3.9.2. (a)

(b) When p has value $\frac{1}{2}$ and q has value 0.

(c) We write Γ for the set of undischarged assumptions of D. The only tricky case is when D has the form

$$
\frac{\begin{array}{c} \not\phi \\ D' \\ \psi \end{array}}{(\phi \to \psi)} \ (\to I)
$$

where D' has conclusion ψ and its undischarged assumptions all lie in $\Gamma \cup \{\phi\}$. Let A be any structure with $A^\star(\phi \to \psi) < 1$, and suppose for contradiction that $A^\star(\chi) > A^\star(\phi \to \psi)$ for all $\chi \in \Gamma$. Since $A^\star(\phi \to \psi) < 1$, we are in one of the cases (i) $A^\star(\phi) = 1$, $A^\star(\psi) = \frac{1}{2}$, (ii) $A^\star(\phi) > 0 = A^\star(\psi)$. In case (i), $A^\star(\phi \to \psi) = 0.5$, so $A^\star(\chi) = 1$ for all $\chi \in \Gamma$, contradicting the inductive assumption on D'. In case (ii) $A^\star(\chi) > A^\star(\psi)$ for all $\chi \in \Gamma \cup \{\phi\}$, again contradicting the inductive assumption.

5.1.1. The bold symbols below are some of the variables:

Putting $\boldsymbol{\theta} = \frac{\pi}{9}$ and $\boldsymbol{c} = \cos \frac{\pi}{9}$, Example 6.3 gives

$$
\cos 3\theta = 4c^3 - 3c.
$$

However, $\cos 3\theta = \cos \frac{\pi}{3} = \frac{1}{2}$. Hence $\frac{1}{2} = 4c^3 - 3c$. In other words, $c = \cos \frac{\pi}{9}$ is a root of the cubic equation

$$
8x^3 - 6x - 1 = 0.
$$

The bold expressions below are definite descriptions:

Putting $\theta = \frac{\pi}{9}$ and $c = \cos \frac{\pi}{9}$, **Example 6.3** gives

$$
\cos 3\theta = 4c^3 - 3c.
$$

However, $\cos 3\theta = \cos \frac{\pi}{3} = \frac{1}{2}$. Hence $\frac{1}{2} = 4c^3 - 3c$. In other words, $c = \cos \frac{\pi}{9}$ is a root of **the cubic equation**

$$8x^3 - 6x - 1 = 0.$$

The underlined expressions below are some of the complex mathematical terms:

Putting $\theta = \frac{\pi}{9}$ and $c = \underline{\cos \frac{\pi}{9}}$, Example 6.3 gives

$$\underline{\cos 3\theta = 4c^3 - 3c}.$$

However, $\cos 3\theta = \cos \frac{\pi}{3} = \frac{1}{2}$. Hence $\frac{1}{2} = 4c^3 - 3c$. In other words, $c = \cos \frac{\pi}{9}$ is a root of the cubic equation

$$\underline{8x^3 - 6x - 1 = 0.}$$

5.1.2. (a) v is free.

(b) x and y are free, but all three occurrences of t are bound by the limit operator.

(c) The only variable is r; its three occurrences are all bound by 'For all'.

(d) Like (c), with 'some' instead of 'all'.

(e) Leibniz's calculus notation is brilliantly designed for calculations but not always easy to analyse logically. The most straightforward analysis of this equation makes y a free variable for a function of x, and z a variable free at both occurrences. But d^2/dx^2 is an operator that binds the x in it (and an implied x in $y(x)$), though the occurrence of x on the right of the formula is free.

5.2.3. (a) We cannot test whether $(0, 2, 3)$ is in the relation, because y occurs twice on the left, and the same number has to be put for y at both occurrences.

(b) The condition for $(0,0,0)$ to be in the relation is that $0+0 = 0+w$. But because w is not on the left, there is nothing to tell us what w is, so we cannot tell whether $(0,0,0)$ is in the relation.

5.2.5. (a) Substitutable. $x^2 - yz$.

(b) Substitutable. z.

(c) Substitutable. $\int_{x^2-yz}^{z} \sin(w)dw$.

(d) Substitutable (since (d) has no free occurrences of y). $\int_{w}^{z} \sin(y)dy$.

(e) Not substitutable (since putting $x^2 - yz$ for y makes the integral bind the x^2).

5.3.1. The lexicon of $\mathrm{LR}(\sigma)$ consists of the symbols in σ together with the twelve symbols '$\neg$', '$\wedge$', '$\vee$', '$\rightarrow$', '$\leftrightarrow$', '$\perp$', '$=$', '$\forall$', '$\exists$', '(', ')' and ','.

5.3.6. (b) Let t be a term of LR. As for Lemma 3.3.3, a proof by induction on the heights of nodes in the parsing tree of t shows:

  (i) Every initial segment of t has depth $\geqslant 0$; the initial segments with depth 0 are t itself and the first symbol of t.

  (ii) If t is $F(t_1, \ldots, t_n)$ where F is a function symbol of arity n and $t_1, \ldots, t_n$ are terms of LR, then the $n-1$ commas shown are exactly the occurrences of commas in t that have depth 1.

5.4.2. (a) t is $\bar{\cdot}(x, \bar{S}(x))$.

  (b) t is $\bar{\cdot}(y, y)$.

  (c) t is $\bar{+}(x, y)$.

  (d) Possible answer: ϕ is $(z = \bar{\cdot}(\bar{0}, z))$.

5.4.6. Below the formulas are left out to save space.

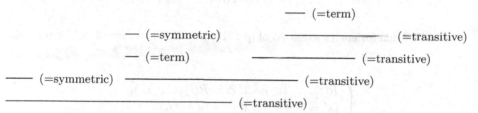

At the left-hand occurrence of (=term), $t[e/v] = t[xz/v]$ is derived from $e = xz$ where t is yv. At the right-hand occurrence of (=term), $s[yx/v] = s[e/v]$ is derived from $yx = e$ where s is vz.

5.4.7. For a cleaner notation we write t' for $t[r/y]$ and ϕ' for $\phi[r/y]$.

Here is the case where a node ν in D carries the right label (=E) (case (e)(v) in Definition 5.4.5). At node ν we have a formula $\phi[t/x]$, and at its daughters we have the formulas $\phi[s/x]$ and $(s = t)$. The variable x here is used purely to mark the site for the substitutions, so we can choose it arbitrarily; assume it is distinct from y and from every variable occurring in D, s or t. The formulas at the corresponding nodes in D' are $\phi[t/x]'$ at ν and $\phi[s/x]'$ and $(s = t)'$ at its daughters. To show that the rule for (=E) is obeyed at ν, we note first that by Definition 5.4.4(c), $(s = t)'$ is $(s' = t')$; so it will suffice to show that $\phi[t/x]'$ is $\phi'[t'/x]$ and $\phi[s/x]'$ is $\phi'[s'/x]$. The argument in both cases is the same. We start by showing that for every term u, $u[t/x]'$ is $u'[t'/x]$, using induction on the complexity of u in Definition 5.4.4. Suppose first that u is y; then since x is not y and is not in t, $u[t/x] = u$, so $u[t/x]' = u' = r = r[t'/x] = u'[t'/x]$. Next suppose that u is x; then

$u[t/x] = t$ so that $u[t/x]' = t' = u[t'/x]$. If u is any other variable then $u[t/x]' = u' = u = u[t'/x] = u'[t'/x]$. Next suppose u is $F(s_1, \ldots, s_n)$ where F is a function symbol and $s_1, \ldots, s_n$ are terms. Then $u[t/x] = F(s_1[t/x], \ldots, s_n[t/x])$, so

$$u[t/x]' = F(s_1[t/x], \ldots, s_n[t/x])' = F(s_1[t/x]', \ldots, s_n[t/x]')$$
$$= F(s_1'[t'/x], \ldots, s_n'[t'/x]) = F(s_1', \ldots, s_n')[t'/x] = u'[t'/x].$$

The cases for $\phi[t/x]'$ are similar, using Definition 5.4.4 throughout.

5.4.8.   Take, for example, the case of eliminating a function symbol F not in ρ. For each term t of $\mathrm{LR}(\sigma)$ we define a term t' by recursion on complexity:

$$t' = \begin{cases} t & \text{if } t \text{ is a variable or constant symbol,} \\ x_0 & \text{if } t \text{ is } F(s_1, \ldots, s_n) \text{ for some terms } s_1, \ldots, s_n, \\ G(s_1', \ldots, s_n') & \text{if } t \text{ is } G(s_1, \ldots, s_n) \text{ with } G \neq F. \end{cases}$$

Then for each formula ϕ of qf $\mathrm{LR}(\sigma)$ we define a formula ϕ' by recursion on complexity:

$$\phi' = \begin{cases} R(t_1', \ldots, t_n') & \text{if } \phi \text{ is } R(t_1, \ldots, t_n), \\ (s' = t') & \text{if } \phi \text{ is } (s = t), \\ \bot & \text{if } \phi \text{ is } \bot, \\ (\psi' \Box \chi') & \text{if } \phi \text{ is } (\psi \Box \chi) \text{ with } \Box \in \{\wedge, \vee, \rightarrow, \leftrightarrow\}, \\ (\neg\psi') & \text{if } \phi \text{ is } (\neg\psi). \end{cases}$$

(Conscientious logicians can prove that t' really is a term and ϕ' really is a formula.) The symbol F never appears in ϕ'; also if ϕ is in qf $\mathrm{LR}(\rho)$ then ϕ' is ϕ. The remainder of the argument is very similar to Exercise 5.4.7.

5.5.2.

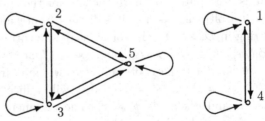

5.5.7.   (a) 5.   (b) 6.   (c) Draw the digraph with edge relation $\{(1,2), (2,3), (3,4)\}$.

5.6.2.   (a) True, (b) false, (c) true, (d) true.

5.6.3.

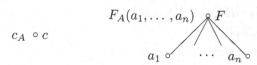

$$c_A \circ c$$

where c is a constant symbol and F is a function symbol of arity n.

5.7.1. (a) $\{0\}$. (b) $\emptyset$.

5.7.2. (c) $\{(4, n) \mid n \in \mathbb{N}\}$. (f) $\{(1, 6), (2, 3), (3, 2), (6, 1)\}$.

5.8.2. (f) There are natural numbers y_1, y_2 such that $y_1(x_1 + 1) = 1 + y_2(x_2 + 1)$. (By Euclid's algorithm there are integers y_1 and y_2 as in this equation. Replacing y_1 by $y_1 + m(x_2 + 1)$ and y_2 by $y_2 + m(x_1 + 1)$ for a large enough m, we can ensure that y_1 and y_2 are natural numbers.)

(g) $x_1 \neq x_2 \Leftrightarrow$ for some y, $y + 1 = (x_1 - x_2)^2$. Multiplying out this equation and applying al-jabr gives: There is a natural number y such that $y + 1 + 2x_1 x_2 = x_1^2 + x_2^2$.

5.10.1. (a) We show, by induction on i, that $F(s_1, \dots, s_n) \sim F(t_1, \dots, t_{i-1}, s_i, \dots, s_n)$ whenever $1 \le i \le n + 1$.
Case 1: $i = 1$. The claim is that $F(s_1, \dots, s_n) \sim F(s_1, \dots, s_n)$, which holds because $\sim$ is reflexive.
Case 2: $i = k + 1$, assuming it for $i = k$. The assumption is that $F(s_1, \dots, s_n) \sim F(t_1, \dots, t_{k-1}, s_k, \dots, s_n)$. Take ϕ to be the formula

$$F(s_1, \dots, s_n) = F(t_1, \dots, t_{k-1}, x, s_{k+1}, \dots, s_n).$$

The assumption states that $\phi[s_k/x]$ is in the Hintikka set Δ. So by clause (7) for Hintikka sets, $\phi[t_k/x]$ is in Δ, in other words, $F(s_1, \dots, s_n) \sim F(t_1, \dots, t_k, s_{k+1}, \dots, s_n)$.
The last step of the induction, when $i = n + 1$, shows that $F(s_1, \dots, s_n) \sim F(t_1, \dots, t_n)$.
(b) is similar but a little simpler.

7.1.1. (f) $\forall x(T(x) \to \exists y \exists z((y \neq z) \land \forall w(L(w, x) \leftrightarrow (w = y \lor w = z))))$.

(g) $\exists x \exists y(T(x) \land T(y) \land x \neq y \land \exists w L(w, x) \land \exists w L(w, y))$.

(h) $\exists x \exists y(T(x) \land T(y) \land x \neq y \land \exists w(L(w, x) \land L(w, y)))$.

7.1.2. $\exists y(P(y) \wedge \exists_{\geqslant n} x(P(x) \wedge x \neq y))$

7.1.3. (a) x is male $\wedge \exists y$ (y is a child of x).

(b) $\exists y$ (y is a child of $x \wedge y$ is female).

7.1.5. Possible answers:

(a) F is an injective function.

(b) F is surjective. (Or: F is onto B.)

(c) F is a bijection from A to B.

(d) F has a fixed point.

(e) A and B are the same set. (Or: $A = B$.)

7.2.4. (b)(i) $FV(\phi) = \{x_1, x_2\}$, $BV(\phi) = \{x_0, x_1\}$.

7.2.5. (a)

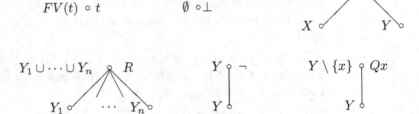

where t is a term, R is a relation symbol of arity n, Q is a
quantifier symbol and x is a variable.

7.2.7. Here is one of the quantifier clauses.

• If x is not y then $\mathrm{Sub}(t, y, \forall x \psi)$ if and only if either (i) $y \notin FV(\psi)$ or
(ii) $\mathrm{Sub}(t, y, \psi)$ and $x \notin FV(t)$.

(This clause is famously easy to get wrong.)

7.3.1. (a)

domain	$\{0, 1\}$;
$\bar{0}$	the number 0;
$S(x)$	x;
$P(x)$	$x = 0$.

7.3.2. (a), (b), (e) and (f) are logically equivalent; (c) and (d) are not logically
equivalent either to (a) or to each other.

7.4.4. The application of ($\forall$I) is faulty. The variable z should not occur in
any assumption which is undischarged at the node carrying $R(z, u)$.
But at that node, $R(z, u)$ is itself an undischarged assumption, since
it becomes discharged only at the bottom node of the diagram. This is
the only error; the other rules, including ($\exists$E), are correctly applied.

7.5.1. (f) False: x^2 is even if and only if x is even.

7.7.3. (a)
$$\forall x E(x, x)$$
$$\forall x \forall y \, (E(x, y) \rightarrow E(y, x))$$
$$\forall x \forall y \forall z \, ((E(x, y) \wedge E(y, z)) \rightarrow E(x, z))$$
$$\forall x \exists y \exists z \, (x \neq y \wedge x \neq z \wedge y \neq z \wedge \forall w \, (E(x, w)$$
$$\leftrightarrow (w = x \vee w = y \vee w = z)))$$

7.7.4. (a) $\forall x(\exists y((y < x) \wedge \forall z((z < x \wedge z \neq y) \rightarrow z < y)) \wedge \exists y((x < y) \wedge \forall z((x < z \wedge y \neq z) \rightarrow y < z)))$.

(b) Each equivalence class is ordered like the ordered set $\mathbb{Z}$ of integers. If we shrink each equivalence class to a single element, the result is again a linearly ordered set. Conversely if J is any linearly ordered set, we construct a linearly ordered set $J \times \mathbb{Z}$ as follows. The elements of the order are the ordered pairs (i, m) where i is an element of J and m is an integer. The order relation on $J \times \mathbb{Z}$ is defined by

$$(i, m) < (j, n) \Leftrightarrow \text{ either } i < j, \text{ or } i = j \text{ and } m < n.$$

This linearly ordered set $J \times \mathbb{Z}$ is discrete without endpoints, and every discrete linearly ordered set without endpoints is a copy of one of this form.

7.8.2. (e) Writing I for the unit circle, let $f : [0, 2\pi] \rightarrow I$ be the function

$$f(\theta) = \begin{cases} e^{i\theta} & \text{if } \theta \neq 1 - 2^{-n} \text{ for all } n \in \mathbb{N}; \\ e^{i(1 - 2^{(-n-1)})} & \text{if } \theta = 1 - 2^{-n}. \end{cases}$$

7.8.5. Define the *height* of the polynomial $a_0 + a_1 x + \cdots + a_n x^n$ over the integers to be $n + |a_0| + |a_1| + \cdots + |a_n|$. There are only finitely many polynomials of given height, and each has only finitely many roots. So the set of algebraic numbers is the union of countably many finite sets, whence countable.

7.9.1. Let ϕ_n be the sentence

$$\exists x_1 \exists x_2 \cdots \exists x_n (x_1 \neq x_2 \wedge \cdots \wedge x_1 \neq x_n \wedge x_2 \neq x_3 \wedge \cdots \wedge x_2 \neq x_n \wedge$$
$$\cdots \wedge x_{n-1} \neq x_n)$$

and let ψ_n be $(\phi_n \wedge (\neg \phi_{n+1}))$. (See also Exercise 7.1.2.)

7.9.6. (d) For each $m \in \mathbb{N}$ the interval $(m, m + 1)$ in $\mathbb{Q}$ is a countable dense linearly ordered set without endpoints. List its elements as $\{x_0, x_1, \dots\}$ and as $\{y_0, y_1, \dots\}$ with $x_0 \neq y_0$, and apply the proof of Theorem 7.9.6 to find an automorphism g_m of this interval with $g_m(x_0) = y_0$. Let h_m be the identity function on the interval. Then g_m and h_m are distinct automorphisms of the interval. Now let F be as in Theorem 7.8.5 but taking values in $\{0, 1\}$, and for each $f \in F$ define a function $k_f : \mathbb{Q} \to \mathbb{Q}$ by

$$k_f(q) = \begin{cases} g_m(q) & \text{if } q \in (m, m+1) \text{ with } m \geqslant 0, \text{ and } f(m) = 0, \\ h_m(q) & \text{if } q \in (m, m+1) \text{ with } m \geqslant 0, \text{ and } f(m) = 1, \\ q & \text{if } q < 0 \text{ or } q \in \mathbb{N}. \end{cases}$$

Then if $f \neq f'$, the functions k_f and $k_{f'}$ are distinct automorphisms of the linearly ordered set of the rationals. Since F is uncountable by Theorem 7.8.5, and the function $f \mapsto k_f$ is injective, the set of automorphisms of the linearly ordered set of the rationals is uncountable too.

Index

Printed in the United States
By Bookmasters